Groundwater Technical Procedures of the U.S. Geological Survey

Compiled by William L. Cunningham and Charles W. Schalk

Techniques and Methods 1–A1

U.S. Department of the Interior
U.S. Geological Survey

U.S. Department of the Interior
KEN SALAZAR, Secretary

U.S. Geological Survey
Marcia K. McNutt, Director

U.S. Geological Survey, Reston, Virginia: 2011

For more information on the USGS—the Federal source for science about the Earth, its natural and living resources, natural hazards, and the environment, visit http://www.usgs.gov or call 1-888-ASK-USGS

For an overview of USGS information products, including maps, imagery, and publications, visit http://www.usgs.gov/pubprod

To order this and other USGS information products, visit http://store.usgs.gov

Suggested citation:
Cunningham, W.L., and Schalk, C.W., comps., 2011, Groundwater technical procedures of the U.S. Geological Survey: U.S. Geological Survey Techniques and Methods 1–A1, 151 p.

Contents

Figures

Tables

Conversion Factors

Inch/Pound to SI

Multiply	By	To obtain
	Length	
inch (in.)	2.54	centimeter (cm)
inch (in.)	25.4	millimeter (mm)
foot (ft)	0.3048	meter (m)
	Volume	
gallon (gal)	3.785	liter (L)
gallon (gal)	0.003785	cubic meter (m^3)
gallon (gal)	3.785	cubic decimeter (dm^3)
cubic foot (ft^3)	28.32	cubic decimeter (dm^3)
cubic foot (ft^3)	0.02832	cubic meter (m^3)
cubic foot (ft^3)	28.32	liter (L)
	Flow rate	
gallon per minute (gal/min)	0.06309	liter per second (L/s)
	Hydraulic conductivity	
foot per day (ft/d)	0.3048	meter per day (m/d)
	Force	
pound (lb)	4.4482	newton ($kg*m/sec^3$)
	Pressure	
pounds per square inch (psi)	0.0689	bars (bar)
pounds per square inch (psi)	703.07	kilograms per square meter (kg/m^3)

Vertical coordinate information is referenced to the North American Vertical Datum of 1988 (NAVD 88).

Horizontal coordinate information is referenced to the North American Datum of 1983 (NAD 83).

Altitude, as used in this report, refers to distance above the vertical datum.

Specific conductance is given in microsiemens per centimeter at 25 degrees Celsius (μS/cm at 25 °C).

Groundwater Technical Procedures of the U.S. Geological Survey

Compiled by William L. Cunningham and Charles W. Schalk

Abstract

A series of groundwater technical procedures documents (GWPDs) has been released by the U.S. Geological Survey, Water-Resources Discipline, for general use by the public. These technical procedures were written in response to the need for standardized technical procedures of many aspects of groundwater science, including site and measuring-point establishment, measurement of water levels, and measurement of well discharge. The techniques are described in the GWPDs in concise language and are accompanied by necessary figures and tables derived from cited manuals, reports, and other documents. Because a goal of this series of procedures is to remain current with the state of the science, and because procedures change over time, this report is released in an online format only. As new procedures are developed and released, they will be linked to this document.

Introduction

This report is a compilation of groundwater technical procedures documents (GWPDs) that describe measurement and data-handling procedures commonly used by the U.S. Geological Survey (USGS). These technical procedures, which were first compiled in 1995 as an internal tool for USGS technicians and hydrologists, have been collected from common techniques cited in USGS reports, USGS internal memoranda, and USGS training programs for many years. Because of the external demand for documentation of these procedures, and the desire to cite them outside of the USGS, they have been reviewed, edited, and compiled in this document. These techniques are a national resource for USGS Water Science Centers and, as such, may not contain sufficient detail for site-specific complexities for other than USGS users. These techniques are provided as the recommended field procedures for USGS Water Science Centers. Individual Centers are encouraged to document modifications that are made to these procedures in project-specific groundwater quality-assurance plans or the Center's groundwater quality-assurance and quality-control plan.

The GWPDs are written in concise language with step-by-step instructions of sufficient detail so that someone with limited experience with the procedure but with a basic understanding of the measurements and general field work can successfully reproduce the procedure unsupervised. The GWPDs do not provide every detail of an individual field task, as the user is expected to have at least nominal field experience. The user also must be cognizant of local regulations on working in and around groundwater wells. State and local ordinances take precedence over any guidance provided in this report. Each GWPD provides an abbreviated list of references if further detail or background information is required. Figures are included where appropriate, and some GWPDs reference other GWPDs. Hypertext links to illustrations, forms, and reports are provided in the body of each document.

Most GWPDs have the following structure:

- Title
- Version
- Purpose
- Materials and Instruments
- Data Accuracy and Limitations
- Advantages
- Disadvantages
- Assumptions
- Instructions
- Data Recording
- References

This report is designed as an online document for use by groundwater hydrologists, technicians, and data managers. The publication of the GWPDs in this format has several benefits:

- It will provide a reference for citation of techniques used during field investigations;

- It will allow hydrologists, technicians, and data managers from outside the USGS to reference techniques used by the USGS;
- It will provide a consistent set of training materials for those new to the routine aspects of groundwater-data collection and handling;
- It will provide an archive for changes in procedures over time as procedures evolve or as tools and equipment become obsolete.
- It will remain current to state-of-the-science techniques.

This report compiles techniques for groundwater-site establishment, well maintenance, water-level measurements, groundwater-discharge measurements, and single-well aquifer tests. It does not document groundwater-quality techniques. These procedures can be found in "U.S. Geological Survey, National Field Manual for the Collection of Water Quality Data." Many of the methods described in the GWPDs are based on United States Office of Water Data Coordination (1977), Garber and Koopman (1968), and Driscoll (1986).

Purpose and Scope

The purpose of this report is to provide a citable document for technical field procedures used by USGS technicians and hydrologists. These procedures have been used by the USGS as guidance for field work, standardization of measurements and other tasks, training of staff, and quality assurance. USGS Water Science Centers can use these procedures as basic guidance and modify them for their circumstances, hydrologic conditions, project objectives, and Center needs. Modifications to these procedures are documented in project-specific groundwater quality-assurance plans or the Center's groundwater quality-assurance and quality-control plan.

The scope of this report generally is restricted to common field-based procedures. Although instrument calibration in the office environment is an integral part of the quality assurance of USGS field work, office-based calibration procedures are not directly addressed in these field procedures. This report does not provide documentation of all procedures used by the Water Science Centers in the USGS, and it does not cover field techniques that are used to meet special objectives. For instance, a USGS project's objectives may require an accuracy and (or) precision not supported by these methods. In those cases, these methods are modified by the individual project and documented in the accompanying project reports.

Review and Revision

GWPDs, like any standard operating procedure, should remain current. The documents will be updated periodically as errors are detected, equipment changes, or new standard techniques evolve. Each procedure is consecutively numbered and contains a version number/date. Those wishing to cite these procedures should include the version number/date of the procedure as an integral part of the reference. These procedures will change with time, and the version number will change accordingly. New procedures will be made available as they are developed, and general electronic announcements will accompany releases of new GWPDs.

Older versions of updated procedures will be archived, as will GWPDs that no longer are used or followed. Hypertext links will be reassigned to the new versions of GWPDs so that the most up-to-date version of the document will be available online.

Technical Procedures

GWPD 1—Measuring water levels by use of a graduated steel tape

GWPD 2—Identifying a minimum set of data elements to establish a groundwater site

GWPD 3—Establishing a permanent measuring point and other reference marks

GWPD 4—Measuring water levels by use of an electric tape

GWPD 5—Documenting the location of a well

GWPD 6—Recognizing and removing debris from a well

GWPD 7—Estimating discharge from a naturally flowing well

GWPD 8—Estimating discharge from a pumped well by use of the trajectory free-fall or jet-flow method

GWPD 9—Recording minimum and maximum water levels

GWPD 10—Measuring discharge from a pumped well by use of a circular orifice weir

GWPD 11—Measuring well depth by use of a graduated steel tape

GWPD 12—Measuring water levels in a flowing well

GWPD 13—Measuring water levels by use of an air line

GWPD 14—Measuring continuous water levels by use of a float-activated recorder

GWPD 15—Obtaining permission to install, maintain, or use a well on private property

GWPD 16—Measuring water levels in wells and piezometers by use of a submersible pressure transducer

GWPD 17—Conducting an instantaneous change in head (slug) test with a mechanical slug and submersible pressure transducer

Acknowledgments

The field procedures described in this report have been compiled from existing USGS reports, various other reference documents, and the technical expertise of the compilers. In addition to the references provided, important source materials include unpublished USGS training and field manuals and technical memoranda from the Office of Groundwater. The following USGS staff (retired) contributed substantially to the contents of this document: Jilann O. Brunett, David C. Dickerman, Linda H. Geiger, and Julia A. Huff. The compilers also appreciate the important contribution by the staff of the USGS Science Publishing Network, including Kay Hedrick, Bonnie Turcott, and Jeffrey Corbett.

References Cited

Driscoll, F.G., 1986, Groundwater and wells (2d ed.): St. Paul, Minnesota, Johnson Filtration Systems, Inc., 1089 p.

Garber, M.S., and Koopman, F.C., 1968, Methods of measuring water levels in deep wells: U.S. Geological Survey Techniques of Water-Resources Investigations, book 8, chap. A1, 23 p.

U.S. Geological Survey, Office of Water Data Coordination, 1977, National handbook of recommended methods for water-data acquisition: Office of Water Data Coordination, Geological Survey, U.S. Department of the Interior, chap. 2, 149 p.

GWPD 1—Measuring water levels by use of a graduated steel tape

VERSION: 2010.1

PURPOSE: To measure the depth to the water surface below land-surface datum using the graduated steel tape (wetted-tape) method.

Materials and Instruments

1. A steel tape graduated in feet, tenths and hundredths of feet. A black tape is preferred to a chromium-plated tape. If a chromium-plated tape is used, paint the back of the tape with a flat black paint to make reading the wetted chalk mark easier. A break-away weight should be attached to a ring on the end of the tape with wire strong enough to hold the weight, but not as strong as the tape, so that if the weight becomes lodged in the well the tape can still be pulled free. The weight should be made of brass, stainless steel, or iron. Lead weights are not acceptable.
2. Blue carpenter's chalk.
3. Clean rag.
4. Pencil or pen, blue or black ink. Strikethrough, date, and initial errors; no erasures.
5. Water-level measurement field form, or handheld computer for data entry.
6. Two wrenches with adjustable jaws or other tools for removing well cap.
7. Cleaning supplies for water-level tapes as described in the National Field Manual (Wilde, 2004).
8. Key for well access.

Data Accuracy and Limitations

1. A graduated steel tape is commonly accurate to 0.01 foot.
2. Most accurate for water levels less than 200 feet below land surface.
3. The steel tape should be calibrated against another acceptable steel tape. An acceptable steel tape is one that is maintained in the office for use only for calibrating steel tapes, and this calibration tape never is used in the field.
4. Oil, ice, or debris may interfere with a water-level measurement.
5. Corrections are necessary for measurements made through angled well casings.
6. When measuring deep water levels (greater than 500 feet), tape expansion and stretch is an additional consideration (Garber and Koopman, 1968).

Advantages

1. The graduated steel tape method is considered to be the most accurate method for measuring water levels in nonflowing wells of moderate depth.
2. Easy to use.
3. Small tape diameter allows access through small ports and provides little interference with pump wiring.

Disadvantages

1. Results may be unreliable if water is dripping into the well or condensing on the well casing.
2. Not recommended for measuring water levels while wells are being pumped.
3. Initial measurement is difficult if estimated water level is not known.

4. Wetted chalk mark may dry before tape is retrieved under hot, dry conditions with large depths to water.

Assumptions

1. An established measuring point (MP) exists and the distance from the MP to land-surface datum (LSD) is known (fig. 1). See GWPD 3 for the technical procedure document on establishing a permanent MP.
2. The MP is clearly marked and described so that a person who has not measured the well will be able to recognize it.
3. For established wells, a water-level measurement taken during the last field visit is available to estimate the length of tape that should be lowered into the well.
4. The black sheen on the steel tape has been dulled so that the tape will retain the chalk.
5. The well is free of obstructions that could affect the plumbness of the steel tape and cause errors in the measurement.
6. The same field method is used for measuring depth below measuring point, or depth relative to vertical datum, but with a different datum correction.
7. The graduated steel tape has been calibrated.

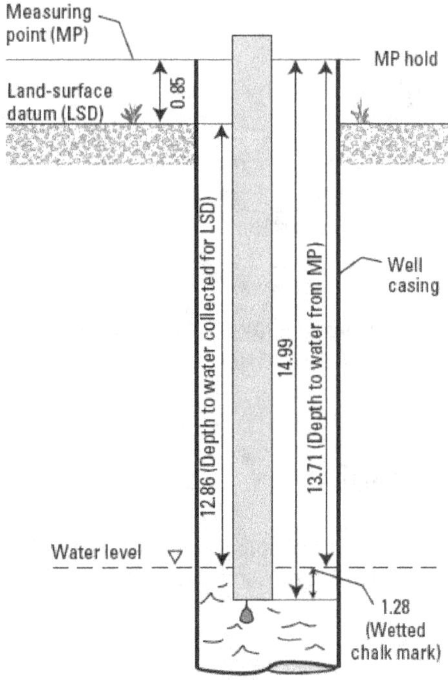

Figure 1. Water-level measurement using a graduated steel tape.

Instructions

1. Open the well.
2. Chalk the lower few feet of the tape by pulling the tape across a piece of blue carpenter's chalk. A wetted chalk mark will identify that part of the tape that was submerged.
3. Review recent measurements from the well, if available, to estimate the hold point on the tape.
4. Refer to figure 1 for an illustration of the elements of a steel tape measurement. Lower the weight and tape into the well until the lower end of the tape is submerged below the water. The weight and tape should be lowered into the water slowly to prevent splashing. Place the thumb and index finger on the tape graduation that is 0.01 less than the next whole foot mark (14.99 in figure 1). Continue to lower the end of the tape into the well until the thumb and index finger meet the MP. Record the graduation value (the HOLD) in the Hold column of the water-level measurement field form (fig. 2).
5. Rapidly bring the tape to the surface before the wetted chalk mark dries and becomes difficult to read. Record the length of the wetted chalk (the CUT) in the Cut row of the water-level measurement field form (fig. 2). Record the time of the measurement in the "Time" row of the form.
6. Subtract the CUT from the HOLD and record this number in the "WL below MP" column of the water-level measurement field form (fig. 2). The difference between the HOLD and the CUT is the depth to water below the MP.
7. If the tape-calibration procedure indicates that a correction is needed at a given water-level depth or for a given water-level range, apply that correction to the "WL below MP" value by adding or subtracting the appropriate correction.
8. Record the MP correction length on the "MP correction" row of the field form (fig. 2); the MP correction is positive if the MP is above land surface and is negative if the MP is below land surface (GWPD 3). Subtract the MP correction from the "WL below MP" value to get the depth to water below or above land-surface datum. Record the water level in the "WL below LSD" column of the water-level measurement field form (fig. 2). If the water level is above LSD, record the depth to water in feet below land surface as a negative number.
9. Make a check measurement by repeating steps 1 through 5. The check measurement should be made using a different HOLD value than that used for the original measurement. If the check measurement does not agree

WATER-LEVEL MEASUREMENT FIELD FORM
Steel Tape Measurement

SITE INFORMATION

SITE ID (C1) ______________

Equipment ID ______________ Date of Field Visit ______________

Station name (C12) ______________

WATER-LEVEL DATA

	1	2	3	4	5
Time					
Hold					
Cut					
Tape correction					
WL below MP					
MP correction					
WL below LSD					

Measured by ______________ COMMENTS* ______________

*Comments should include quality concerns and changes in: M.P., ownership, access, locks, dogs, measuring problems, et al.

MEASURING POINT DATA (for MP Changes)

M.P. REMARKS (C324)	BEGINNING DATE (C321)	ENDING DATE (C322)	M.P. HEIGHT (C323) NOTE: (-) for MP below land surface
______________	month - day - year	- -	.
______________	- -	- -	.

Final Measurement for GWSI

WATER LEVEL TYPE CODE (C243): L (below land surface), M (below meas. pt.), S (sea level)

DATE WATER LEVEL MEASURED (C235)	TIME (C709)	STATUS (C238)	METHOD (C239)	TYPE (C243)	WATER LEVEL (C237)
month - day - year					.

METHOD OF WATER-LEVEL MEASUREMENT(C239)

A	B	C	E	G	H	L	M	N	R	S (GWPD1)	T	V (GWPD4)	Z
airline,	analog,	calibrated airline,	estimated,	pressure gage,	calibrated press. gage,	geophysical logs,	manometer,	non-rec. gage,	reported,	steel tape,	electric tape,	calibrated elec. tape	other

SITE STATUS FOR WATER LEVEL (C238)

D	E	F	G	H	I	J	M	N	O	P	R	S	T	V	W	X	Z	BLANK
dry,	recently flowing,	flowing,	nearby flowing	nearby recently flowing,	injector site,	injector site monitor,	plugged,	measurement discon.,	obstruction,	pumping,	recently pumped,	nearby pumping,	nearby recently pumped,	foreign substance,	well destroyed,	surface water effects,	other	static

Figure 2. Water-level measurement field form for steel tape measurements. This form, or an equivalent custom-designed form, should be used to record field measurements.

with the original measurement within 0.02 foot, continue to make measurements until the reason for lack of agreement is determined or the results are shown to be reliable. If more than two measurements are made, use best judgment to select the measurement most representative of field conditions.

10. Complete the "Final Measurement for GWSI" portion of the field form (fig. 2).
11. After completing the water-level measurement, disinfect and rinse that part of the tape that was submerged below the water surface, as described in the National Field Manual (Wilde, 2004). This will reduce the possibility of contamination of other wells from the tape.
12. Close the well.
13. Maintain the tape in good working condition by periodically checking the tape for rust, breaks, kinks, and possible stretch due to the suspended weight of the tape and the tape weight. The tape should be recalibrated annually and recorded in the calibration logbook.
14. In some pumped wells, a layer of oil may float on the water surface. If the oil layer is a foot or less thick, read the tape at the top of the oil mark and use this value for the water-level measurement instead of the wetted chalk mark. The measurement will differ slightly from the water level that would be measured were the oil not present. However, if several feet of oil are present in the well, or if it is necessary to know the thickness of the oil layer, an electronic "interface probe," or a commercially available water-detector paste can be used that will detect the presence of water in the oil. The paste is applied to the lower end of the tape and will show the top of the oil as a wet line, and the top of the water will show as a distinct color change. Because oil density is about three-quarters that of water, the water level can be estimated by adding the thickness of the oil layer times its density to the oil-water interface altitude.

Data Recording

All calibration and maintenance data associated with steel tape use are recorded in the calibration and maintenance equipment logbook.

All water-level data are recorded on the water-level measurement field form (fig. 2) or by using a handheld computer program such as MONKES. Field measurements are recorded to the nearest 0.01 foot or to the appropriate precision based on the judgment of the hydrographer. When using a handheld computer to record field measurements, the measurement procedure is the same as described in the "Instructions" section.

References

Cunningham, W.L., and Schalk, C.W., comps., 2011, Groundwater technical procedures of the U.S. Geological Survey, GWPD 3—Establishing a permanent measuring point and other reference marks: U.S. Geological Survey Techniques and Methods 1–A1, 13 p.

Garber, M.S., and Koopman, F.C., 1968, Methods of measuring water levels in deep wells: U.S. Geological Survey Techniques of Water-Resources Investigations, book 8, chap. A1, 23 p.

Hoopes, B.C., ed., 2004, User's manual for the National Water Information System of the U.S. Geological Survey, Ground-Water Site-Inventory System (version 4.4): U.S. Geological Survey Open-File Report 2005–1251, 274 p.

Katz, B.G., and Jelinski, J.C., 1999, Replacement materials for lead weights used in measuring ground-water levels: U.S. Geological Survey Open-File Report 99–52, 13 p.

U.S. Geological Survey, Office of Water Data Coordination, 1977, National handbook of recommended methods for water-data acquisition: Office of Water Data Coordination, Geological Survey, U.S. Department of the Interior, chap. 2, 149 p.

Wilde, F.D., ed., 2004, Cleaning of equipment for water sampling (version 2.0): U.S. Geological Survey Techniques of Water-Resources Investigations, book 9, chap. A3, accessed July 17, 2006, at *http://pubs.water.usgs.gov/twri9A3/*.

GWPD 2—Identifying a minimum set of data elements to establish a groundwater site

VERSION: 2010.1

PURPOSE: To specify the minimum amount of information that should be collected during the initial site inventory in the field for an individual groundwater site. These data will be recorded in the National Water Information System (NWIS).

Materials and Instruments

1. Best available paper maps or Global Positioning System (GPS) receiver
2. Groundwater Site Inventory (GWSI) System Groundwater Site Schedule, Form 9-1904-A
3. Spray paint, bright color
4. Metal file for marking well casing; hammer and cold steel chisel, survey monument (nail, spike, tablet)
5. Camera
6. Protractor, calculator, or other tools to calculate angles and lengths
7. Rod, leveling instrument, and leveling notes sheets
8. A steel tape graduated in feet, tenths and hundredths of feet
9. Blue carpenter's chalk
10. Clean rag
11. Field notebook
12. Pencil or pen, blue or black ink. Strikethrough, date, and initial errors; no erasures
13. Water-level measurement field form, or handheld computer for data entry
14. Two wrenches with adjustable jaws or other tools for removing well cap
15. Cleaning supplies for water-level tapes as described in the National Field Manual (Wilde, 2004)
16. Key for well access

Data Accuracy and Limitations

1. Altitudes determined from topographic maps are accurate to within one-half the map contour interval; latitudes and longitudes are accurate to about 0.5 second.
2. Accuracy of latitude, longitude, and altitudes determined by use of GPS are dependent on each instrument's capabilities.
3. The accuracy of the measuring point, land-surface datum, measuring point correction, and reference marks depends on the measurement method used. See GWPD 3 for additional information.
4. A graduated steel or electric tape commonly is accurate to 0.01 foot. See GWPD 1 and GWPD 4 for additional information.

Assumptions

1. The groundwater site is established by a field visit. At times, a site is established without a field visit. In that instance, less information may be available to establish the site in GWSI.
2. A groundwater site is a single point, not a geographic area or property.
3. All information available for a site will be compiled and entered in GWSI. This includes data and information that are not mandatory for GWSI (*http://nwis.usgs.gov/nwisdocs4_10/gw/gwintrocoding_Sect2-0.pdf*).
4. A GPS unit and (or) paper maps will be used to complete the location-based information needed for Form 9-1904-A (fig. 1). A U.S. Geological Survey (USGS) computer

application is available for this task which automates some of the steps in this procedure. Use of that application is encouraged, but it is not yet available for field use.

5. The hydrographer has gathered all of the information available about the well, including a well-construction log, geologic log, owner information, and has permission to access the well.

Instructions

1. Locate the well as described in GWPD 5.
2. Establish a permanent measuring point, land-surface datum, and nearby reference marks as described in GWPD 3.
3. Measure the total depth of the well, as described in GWPD 11.
4. Measure the water level in the well, as described in GWPD 1 or GWPD 4.
5. Use the information collected prior to the field visit and the measurements collected during the field visit to complete every GWSI component (fig. 1) for which you have information.

Data Recording

Data are recorded in the field on the GWSI Groundwater Site Schedule (Form 9-1904-A, fig. 1). Water levels also are recorded on the appropriate water-level measurement field form.

References

American Society for Testing and Materials, 1994, ASTM standards on ground water and vadose zone investigations (2d ed.): Philadelphia, Pennsylvania, American Society for Testing and Materials, p. 300–304.

Cunningham, W.L., and Schalk, C.W., comps., 2011a, Groundwater technical procedures of the U.S. Geological Survey, GWPD 1—Measuring water levels by use of a graduated steel tape: U.S. Geological Survey Techniques and Methods 1–A1, 4 p.

Cunningham, W.L., and Schalk, C.W., comps., 2011b, Groundwater technical procedures of the U.S. Geological Survey, GWPD 3—Establishing a permanent measuring point and other reference marks: U.S. Geological Survey Techniques and Methods 1–A1, 13 p.

Cunningham, W.L., and Schalk, C.W., comps., 2011c, Groundwater technical procedures of the U.S. Geological Survey, GWPD 4—Measuring water levels by use of an electric tape: U.S. Geological Survey Techniques and Methods 1–A1, 6 p.

Cunningham, W.L., and Schalk, C.W., comps., 2011d, Groundwater technical procedures of the U.S. Geological Survey, GWPD 5—Documenting the location of a well: U.S. Geological Survey Techniques and Methods 1–A1, 10 p.

Cunningham, W.L., and Schalk, C.W., comps., 2011e, Groundwater technical procedures of the U.S. Geological Survey, GWPD 11—Measuring well depth by use of a graduated steel tape: U.S. Geological Survey Techniques and Methods 1–A1, 10 p.

Hoopes, B.C., ed., 2004, User's manual for the National Water Information System of the U.S. Geological Survey, Ground-Water Site-Inventory System (version 4.4): U.S. Geological Survey Open-File Report 2005–1251, 274 p.

Wilde, F.D., ed., 2004, Cleaning of equipment for water sampling (version 2.0): U.S. Geological Survey Techniques of Water-Resources Investigations, book 9, chap. A3, accessed July 17, 2006, at *http://pubs.water.usgs.gov/twri9A3/*.

FORM NO. 9-1904-A
Revised Sept 2009, NWIS 4 9

File Code ____________
Date ____________

Coded by ____________
Checked by ____________
Entered by ____________

U.S DEPT. OF THE INTERIOR
GEOLOGICAL SURVEY

GROUNDWATER SITE SCHEDULE
General Site Data

AGENCY CODE (C4) USGS | SITE D (C1) | PROJECT (C5)

STATION NAME (C12/900)

1 SITE TYPE (C802) Primary - Secondary | DISTRICT (C6) | COUNTRY (C41) | STATE (C7)

COUNTY or TOWN (C8) | County code

LATITUDE (C9) | LONGITUDE (C10) | LAT/LONG ACCURACY (C11): H Hndrth sec. | 1 tenth sec. | 5 half sec. | S sec. | R 3 sec. | F 5 sec. | T 10 sec. | M min. | U Un-known

LAT/LONG METHOD (C35): C land net | D DGPS | G GPS | L LORAN | M map | N inter-polated digital map | R reported | S survey | U un-known

LAT/LONG DATUM (C36): NAD27 North American Datum of 1927 | NAD83 North American Datum of 1983

ALTITUDE (C16)

ALTITUDE ACCURACY (C18)

ALTITUDE METHOD (C17): A altimeter | D DGPS | G GPS | I IfSAR | J LIDAR | L Level | M map | N DEM | R re-ported | U un-known

ALTITUDE DATUM (C22): NGVD29 National Geodetic Vertical Datum of 1929 | NAVD88 North American Vertical Datum of 1988

LAND NET (C13): ¼ ¼ ¼ S section T township range merid

TOPO-GRAPHIC SETTING (C19): A alluvial fan | B playa | C stream channel | D depres-sion | E dunes | F flat | G flood-plain | H hill-top | K sink-hole | L lake or swamp | M mangrove swamp | O off-shore | P pedi-ment | S hill-side | T ter-race | U undu-lating | V valley flat | W upland draw

HYDROLOGIC UNIT CODE (C20) | DRA NAGE BASIN CODE (C801) | STANDARD T ME ZONE (C813) | DAYLIGHT SAVINGS T ME FLAG (C814) Y OR **N**

MAP NAME (C14) | MAP SCALE (C15)

AGENCY USE (C803): A active no/na | D discon-tinued | I inactive site | L active written | M active oral | O inventory site | R remediated

2 NATIONAL WATER-USE (C39)

DATA TYPE (C804)
Place an 'A' (active), an 'I' (inactive), or an 'O' (inventory) in the appropriate box

WL cont | WL int | QW cont | QW int | PR cont | PR int | EV cont | EV int | wind vel. | tide cont | tide int | sed. con | sed. ps | peak flow | low flow | state water use

INSTRUMENTS (C805)
(Place a "Y" in the appropriate box):

digital reo-order | graphic reo-order | tele-metry land line | tele-metry radio | tele-metry satellite | AHDAS | crest-stage gage | tide gage | deflec-tion meter | bubble gage | stilling well | CR type recorder | weigh-ing rain gage | tipping bucket rain gage | acoustic velocity meter | electro-magnetic flowmeter | pressure transducer

DATE NVENTORIED (C711): month – day – year

RECORD READY FOR WEB (C32): Y ready to display | C condi-tional | P proprie-tary | L local use only

REMARKS (C806)

FOOTNOTES

1SITE TYPE (C802)

GL	Glacier	OC	Ocean	GW	Well	SB	Subsurface
WE	Wetland	OC-CO	Coastal	GW-CR	Collector or Ranney type well	SB-CV	Cave
AT	Atmosphere	LK	Lake, Reservoir, Impoundment	GW-EX	Extensometer well	SB-GWD	Groundwater drain
ES	Estuary			GW-HZ	Hyporheic -zone well	SB-TSM	Tunnel, shaft, or mine
LA	Land	SP	Spring	GW-IW	Interconnected wells	SB-UZ	Unsaturated zone
LA-EX	Excavation	ST	Stream	GW-TH	Test hole not completed as a well		
LA-OU	Outcrop	ST-CA	Canal	GW-MW	Multiple wells		
LA-SNK	Sinkhole	ST-DCH	Ditch				
LA-SH	Soil hole	ST-TS	Tidal strea m				
LA-SR	Shore	FA-WIW	Waste-Injection well				

2 WS water supply | DO domestic | CO commer-cial | IN industrial | IR irrigation | MI mining | LV livestock | PH power hydro-electric | ST waste water treatment | RM remedia-tion | TE thermo-electric power | AQ aqua-culture

C22 Other (see manual for codes)
C36 Other (see manual for codes)
C39 is mandatory for all sites having data in SWUDS.

Figure 1. Groundwater Site Schedule, Form 9-1904-A.

GENERAL SITE DATA

DATA RELIAB LITY (C3): C field checked | L poor location | M minimal data | U unchecked

DATE OF F RST CONSTRUCTION (C21): month – day – year

USE OF SITE (C23): A anode | C standby emer. supply | D drain | E geo-thermal | G seismic | H heat reservoir | M mine | O obser-vation | P oil or gas | R recharge | S repres-surize | T test | U unused | V with-drawal/ return | W with-drawal | X waste | Z des-troyed

SECOND-ARY USE OF SITE (C301) (See use of site)

TERTIARY USE OF SITE (C302) (See use of site)

USE OF WATER (C24): A air cond. | B bottling | C comm-ercial | D de-water | E power | F fire | H domes-tic | I irri-gation | J indus-trial (cooling) | K mining | M medi-cinal | N indus-trial | P public supply | Q aqua-culture | R recrea-tions | S stock | T insti-tutional | U unused | Y desalin-ation | Z other

SECOND-ARY USE OF WATER (C25) (see use of water)

TERTIARY USE OF WATER (C26) (see use of water)

AQUIFER TYPE (C713): U unconfined single | N unconfined multiple | C confined single | M confined multiple | X mixed

PR MARY AQUIFER (C714)

NATIONAL AQUIFER (C715)

HOLE DEPTH (C27)

WELL DEPTH (C28)

SOURCE OF DEPTH DATA (C29): A other gov't | D driller | G geol-ogist | L logs | M memory | O owner | R other reported | S reporting agency | Z other

WATER-LEVEL DATA

DATE WATER-LEVEL MEASURED (C235): month – day – year

T ME (C709)

WATER-LEVEL TYPE CODE (C243): L land surface | M meas. pt. | S vertical datum

WATER LEVEL (C237/241/242)

MP SEQUENCE NO. (C248) (Mandatory if WL type=M)

WATER-LEVEL DATUM (C245) (Mandatory if WL type=S): NGVD29 National Geodetic Vertical Datum Of 1929 | NAVD88 North American Vertical Datum Of 1988 | Other (See manual for codes)

SITE STATUS FOR WATER LEVEL (C238): A atmos. pressure | B tide stage | C ice | D dry | E recently flowing | F flowing | G nearby flowing | H nearby recently flowing | I injector site | J injector site monitor | M plugged | N measure-ment discontinued | O obstruc-tion | P pumping | R recently pumped | S nearby pumping | T nearby recently pumped | V foreign sub-stance | W well des-troyed | X affected by surface water | Z other

METHOD OF WATER-LEVEL MEASUREMENT(C239): A airline | B analog | C calibrated airline | D differ-ential GPS | E esti-mated | F trans-ducer | G pressure gage | H calibrated press. gage | L geophysi-cal logs | M mano-meter | N non-rec. gage | O observed | P acoustic pulse | R reported | S steel tape | T electric tape | V calibrated elec. tape | Z other

WATER-LEVEL ACCURACY (C276): 0 foot | 1 tenth | 2 hun-dredth | 9 not to nearest foot

SOURCE OF WATER-LEVEL DATA (C244): A other gov't | D driller's log | G geol-ogist | L geophysi-cal logs | M memory | O owner | R other reported | S reporting agency | Z other

PERSON MAK NG MEASUREMENT (C246) (WATER LEVEL PARTY)

MEASURING AGENCY (C247) (SOURCE)

EQUIP D (C249) (20 char)

REMARKS (C267) (256 char)

RECORD READY FOR WEB (C858): Y ready to display | C condi-tional | P proprie-tary | L local use only

CONSTRUCTION DATA

RECORD TYPE (C754): CONS

RECORD SEQUENCE NO. (C723)

DATE OF COMPLETED CONSTRUCTION (C60): month – day – year

NAME OF CONTRACTOR (C63)

SOURCE OF DATA (C64): A other gov't | D driller | G geol-ogist | L logs | M memory | O owner | R other reported | S reporting agency | Z other

METHOD OF CONSTRUCTION (C65): A air-rotary | B bored or augered | C cable tool | D dug | H hydraulic rotary | J jetted | P air per-cussion | R reverse rotary | S sonic | T trenching | V driven | W drive wash | Z other

TYPE OF F NISH (C66): C porous concrete | F gravel w/perf. | G gravel screen | H horiz. gallery | O open end | P perf or slotted | S screen | T sand point | W walled | X open hole | Z other

TYPE OF SEAL (C67): B bentonite | C clay | G cement grout | N none | Z other

BOTTOM OF SEAL (C68)

METHOD OF DEVELOPMENT (C69): A air-lift pump | B bailed | C compres-sed air | J jetted | N none | P pumped | S surged | Z other

HOURS OF DEVELOPMENT (C70)

SPECIAL TREATMENT (C71): C chem-icals | D dry ice | E explo-sives | F defloc-culent | H hydro-frac-turing | M mech-anical | Z other

2 - Groundwater Site Schedule

CONSTRUCTION HOLE DATA (3 sets shown)

RECORD TYPE (C756) HOLE   RECORD SEQUENCE NO. (C724)   SEQUENCE NO. OF PARENT RECORD (C59)

DEPTH TO TOP OF INTERVAL (C73)   DEPTH TO BOTTOM OF INTERVAL (C74)   DIAMETER OF INTERVAL (C75)

RECORD SEQUENCE NO. (C724)

DEPTH TO TOP OF INTERVAL (C73)   DEPTH TO BOTTOM OF INTERVAL (C74)   DIAMETER OF INTERVAL (C75)

RECORD SEQUENCE NO. (C724)

DEPTH TO TOP OF INTERVAL (C73)   DEPTH TO BOTTOM OF INTERVAL (C74)   DIAMETER OF INTERVAL (C75)

CONSTRUCTION CASING DATA (4 sets shown)

RECORD TYPE (C758) CSNG   RECORD SEQUENCE NO. (C725)   SEQUENCE NO. OF PARENT RECORD (C59)

DEPTH TO TOP OF CASING (C77)   DEPTH TO BOTTOM OF CASING (C78)   DIAMETER OF CASING (C79)

4 CASING MATERIAL (C80)   CASING THICKNESS (C81)

RECORD SEQUENCE NO. (C725)   SEQUENCE NO. OF PARENT RECORD (C59)

DEPTH TO TOP OF CAS NG (C77)   DEPTH TO BOTTOM OF CAS NG (C78)   DIAMETER OF CASING (C79)

4 CASING MATERIAL (C80)   CASING THICKNESS (C81)

RECORD SEQUENCE NO. (C725)   SEQUENCE NO. OF PARENT RECORD (C59)

DEPTH TO TOP OF CASING (C77)   DEPTH TO BOTTOM OF CASING (C78)   DIAMETER OF CASING (C79)

4 CASING MATERIAL (C80)   CASING THICKNESS (C81)

RECORD SEQUENCE NO. (C725)   SEQUENCE NO. OF PARENT RECORD (C59)

DEPTH TO TOP OF CASING (C77)   DEPTH TO BOTTOM OF CASING (C78)   DIAMETER OF CASING (C79)

4 CASING MATERIAL (C80)   CASING THICKNESS (C81)

FOOTNOTE:

[4] CASING MATERIAL CODES

A	B	C	D	E	F	G	H	I	J	K	L	M	N	P	Q	R	S	T	U	V	W	X	Y	Z	4	6
abs	brick	concrete	copper	PTFE	Fiber-glass	galv. iron	Fiber-glass plastic	wrought iron	Fiber-glass epoxy	PVC thread-ed	glass	other metal	PVC glued	PVC or plastic	FEP	rock or stone	steel	tile	coated steel	stain-less steel	wood	steel carbon	steel galva-nized	other mat.	stain-less 304	stain-less 316

CONSTRUCTION OPENINGS DATA (3 sets shown)

RECORD TYPE (C760) OPEN RECORD SEQUENCE NO. (C726) SEQUENCE NO. OF PARENT RECORD (C59)

DEPTH TO TOP OF INTERVAL (C83) DEPTH TO BOTTOM OF INTERVAL (C84) DIAMETER OF INTERVAL (C87)

[5] MATERIAL TYPE (C86) [6] TYPE OF OPENING (C85) LENGTH OF OPENING (C89) WIDTH OF OPENING (C88)

RECORD SEQUENCE NO. (C726)

DEPTH TO TOP OF INTERVAL (C83) DEPTH TO BOTTOM OF INTERVAL (C84) DIAMETER OF INTERVAL (C87)

[5] MATERIAL TYPE (C86) [6] TYPE OF OPENING (C85) LENGTH OF OPENING (C89) WIDTH OF OPENING (C88)

RECORD SEQUENCE NO. (C726)

DEPTH TO TOP OF INTERVAL (C83) DEPTH TO BOTTOM OF INTERVAL (C84) DIAMETER OF INTERVAL (C87)

[5] MATERIAL TYPE (C86) [6] TYPE OF OPENING (C85) LENGTH OF OPENING (C89) WIDTH OF OPENING (C88)

FOOTNOTES:

[5] TYPE OF MATERIAL CODES FOR OPEN SECTIONS

A	B	C	D	E	F	G	H	I	J	K	L	M	N	P	Q	R	S	T	V	W	X	Y	Z	4	6
ABS	brass or bronze	concrete	ceramic	PTFE	fiber-glass	galv. iron	fiber-glass plastic	wrought iron	fiber-glass epoxy	PVC thread-ed	glass	other metal	PVC glued	PVC	FEP	stain-less steel	steel	tile	brick	mem-brane	steel carbon	steel galva-nized	other	stain-less 304	stain-less 316

[6] TYPE OF OPENINGS CODES

F	L	M	P	R	S	T	W	X	Z
fractured rock	louvered or shutter-type	mesh screen	perforated, porous or slotted	wire-wound screen	screen (unk.)	sand point screen	walled or shored	open hole	other

CONSTRUCTION MEASURING POINT DATA

RECORD TYPE (C766) MPNT RECORD SEQUENCE NO. (C728) BEGINNING DATE (C321) month – day – year ENDING DATE (C322)

M.P. HEIGHT (C323) ALTITUDE OF MEASURING POINT (C325) ALTITUDE METHOD (C326) ALTITUDE ACCURACY (C327)

ALTITUDE DATUM (C328) M.P. REMARKS (C324)

RECORD READY FOR WEB (C857)

Y	C	P	L
ready to display	condi-tional	proprie-tary	local use only

CONSTRUCTION LIFT DATA

RECORD TYPE (C752) LIFT RECORD SEQUENCE NO. (C254) TYPE OF LIFT (C43)

A	B	C	J	P	R	S	T	U	X	Z
air	bucket	centri-fugal	jet	piston	rotary	submer-sible	turbine	un-known	no lift	other

DATE RECORDED (C38) month – day – year

PUMP INTAKE DEPTH (C44)

TYPE OF POWER (C45)

D	E	G	H	L	N	S	W	Z
diesel	electric	gaso-line	hand	LP gas	natural gas	solar	windmill	other

HORSE-POWER RATING (C46) .

MANUFACTURER (C48)

SERIAL NO. (C49)

POWER COMPANY (C50)

POWER COMPANY ACCOUNT NUMBER (C51)

POWER METER NUMBER (C52)

PUMP RATING (C53) (million gallons/units of fuel) .

ADDITIONAL LIFT (C255)

PERSON OR COMPANY MAINTAINING PUMP (C54)

RATED PUMP CAPACITY (gpm) (C268)

STANDBY POWER (C56) (see TYPE OF POWER)

HORSEPOWER OF STANDBY POWER SOURCE (C57) .

MISCELLANEOUS OWNER DATA

RECORD TYPE (C768) OWNR RECORD SEQUENCE NO. (C718) DATE OF OWNERSH P (C159) – –

WU OWNER TYPE (C350)

CP	GV	IN	MI	OT	TG	WS
Corporation	Govern-ment	Individual	Military	Other	Tribal	Water Supplier

END DATE OF OWNERSHIP (C374) – –

OWNER'S NAME (C161)

EXAMPLES: JONES, RALPH A.
JONES CONSTRUCTION COMPANY

OWNER'S PHONE NUMBER (C351)

ACCESS TO OWNER'S NAME (C352)

0	1	2	3	4
Public Access	Coop-erator	USGS Only	District Only	Proprietary

OWNER'S ADDRESS (LINE 1) (C353)

OWNER'S ADDRESS (LINE 2) (C354)

OWNER'S CITY NAME (C355)

STATE (C356)

OWNER'S Z P CODE (C357) –

OWNER'S COUNTRY NAME (C358)

ACCESS TO OWNER'S PHONE/ADDRESS (C359)

0	1	2	3	4
Public Access	Coop-erator	USGS Only	District Only	Proprietary

MISCELLANEOUS VISIT DATA

RECORD TYPE (C774) VIST RECORD SEQUENCE NO. (C737) DATE OF VISIT (C187) month – day – year

NAME OF PERSON (C188)

MISCELLANEOUS OTHER ID DATA (2 sets shown)

RECORD TYPE (C770) O T I D

RECORD SEQUENCE NO. (C736)

OTHER ID (C190)

ASSIGNER (C191)

RECORD SEQUENCE NO. (C736)

OTHER ID (C190)

ASSIGNER (C191)

MISCELLANEOUS OTHER DATA

RECORD TYPE (C772) O T D T

RECORD SEQUENCE NO. (C312)

OTHER DATA TYPE (C181)

OTHER DATA LOCATION (C182) C Cooperator's Office, D District Office R Reporting Agency Z other

DATA FORMAT (C261) F files, M machine readable, P published, Z other

MISCELLANEOUS LOGS DATA (3 sets shown)

RECORD TYPE (C778) L O G S

RECORD SEQUENCE NO. (C739)

TYPE OF LOG (C199)

BEG NN NG DEPTH (C200) .

ENDING DEPTH (C201) .

SOURCE OF DATA (C202) A other gov't D driller G geol-ogist L logs M memory O owner R other reported S reporting agency Z other

DATA FORMAT (C225) F files M machine readable P published Z other

OTHER DATA LOCATION (C226)

RECORD TYPE (C778) L O G S

RECORD SEQUENCE NO. (C739)

TYPE OF LOG (C199)

BEG NNING DEPTH (C200) .

ENDING DEPTH (C201) .

SOURCE OF DATA (C202) A other gov't D driller G geol-ogist L logs M memory O owner R other reported S reporting agency Z other

DATA FORMAT (C225) F files M machine readable P published Z other

OTHER DATA LOCATION (C226)

RECORD TYPE (C778) L O G S

RECORD SEQUENCE NO. (C739)

TYPE OF LOG (C199)

BEGINNING DEPTH (C200) .

ENDING DEPTH (C201) .

SOURCE OF DATA (C202) A other gov't D driller G geol-ogist L logs M memory O owner R other reported S reporting agency Z other

DATA FORMAT (C225) F files M machine readable P published Z other

OTHER DATA LOCATION (C226)

ACOUSTIC LOG:
AS Sonic
AV Acoustic velocity
AW Acoustic waveform
AT Acous ic televiewer

CALIPER LOG:
CP Caliper
CS Caliper, single arm
CT Caliper, three arm
CM Caliper, multi arm
CA Caliper, acous ic

DRILLING LOG:
DT Drilling ime
DR Drillers
DG Geologists
DC Core

ELECTRIC LOG:
EE Electric
ER Single-point resistance
EP Spontaneous potential
EL Long-normal resistivity
ES Short-normal resis ivity
EF Focused resistivity
ET Lateral resistivity
EN Microresistivity
EC Microresistivity, forused
EO Microresis ivity, lateral
ED Dipmeter

ELECTROMAGNETIC LOG:
MM Magnetic log
MS Magnetic susceptibiity log
MI Electromagne ic induc ion log
MD Electromagne ic dual induction log
MR Radar reflection image log
MV Radar direct-wave velocity log
MA Radar direct-wave amplitude log

FLUID LOG:
FC Fluid conductivity
FR Fluid resistivity
FT Fluid temperature
FF Fluid differen ial temperature
FV Fluid velocity
FS Spinner flowmeter
FH Heat-pulse flowmeter
FE Electromagnetic flowmeter
FD Doppler flowmeter
FA Radioactive tracer
FY Dye tracer
FB Brine tracer

NUCLEAR LOG:
NG Gamma
NS Spectral gamma
NA Gamma-gamma
NN Neutron
NT Neutron activitation
NM Neuclear magnetic resonance

OPTICAL LOG:
OV Video
OF Fisheye video
OS Sidewall video
OT Optical televiewer

COMBINATION LOG:
ZF Gamma, fluid resistivity, temperature
ZI Gamma, electromagne ic induction
ZR Long/short normal resistivity
ZT Fluid resis ivity, temperature
ZM Electromagnetic flowmeter, fluid resistivity, temperature
ZN Long/short normal resistivity, spontaneous poten ial
ZP Single-point resistance, spontaneous potential
ZE Gamma, long/short normal resistivity, spontaneous potential, single-point resistance, fluid resi ivity, temperature

WELL CONSTRUCTION LOG:
WC Casing collar
WD Borehole deviation

OTHER LOG:
OR Other

6 - Groundwater Site Schedule

MISCELLANEOUS NETWORK DATA (3 types shown)

RECORD TYPE (C780) N E T W — RECORD SEQUENCE NO. (C730) — TYPE OF NETWORK (C706) Q W water quality — BEGINNING YEAR (C115) — ENDING YEAR (C116)

TYPE OF ANALYSIS (C120)

A	B	C	D	E	F	G	H	I	J	K	L	M	N	P	Z
physical proper-ties	common ions	trace elements	pesti-cides	nutri-ents	sanitary analysis	codes D&B	codes B&E	codes B&C	codes B&F	codes D&E	codes C,D&E	all or most	codes B&C& radio-active	codes B,C&A	other

SOURCE AGENCY (C117) — [7] FREQUENCY OF COLLECTION (C118) — ANALYZING AGENCY (C307) — [8] PRIMARY NETWORK SITE (C257) — [8] SECONDARY NETWORK SITE (C708)

RECORD TYPE (C780) N E T W — RECORD SEQUENCE NO. (C730) — TYPE OF NETWORK (C706) W L water level — BEGINNING YEAR (C115) — ENDING YEAR (C116)

SOURCE AGENCY (C117) — [7] FREQUENCY OF COLLECTION (C118) — [8] PRIMARY NETWORK SITE (C257) — [8] SECONDARY NETWORK SITE (C708)

RECORD TYPE (C780) N E T W — RECORD SEQUENCE NO. (C730) — TYPE OF NETWORK (C706) W D pumpage or with-drawals — BEGINNING YEAR (C115) — ENDING YEAR (C116)

SOURCE AGENCY (C117) — [7] FREQUENCY OF COLLECTION (C118) — METHOD OF COLLECTION (C133) — [8] PRIMARY NETWORK SITE (C257) — [8] SECONDARY NETWORK SITE (C708)

C	E	M	U	Z
calcu-lated	esti-mated	meter-ed	un-known	other

FOOTNOTES:

[7] FREQUENCY OF COLLECTION CODES

A	B	C	D	F	I	M	O	Q	S	W	Z	2	3	4	5	X
annually	bi monthly	continu-ously	daily	semi-monthly	inter mittent	monthly	one-time only	quarter-ly	semi-annually	weekly	other	bi-annually	every 3 years	every 4 years	every 5 years	every 10 years

[8] NETWORK SITE CODES

1	2	3	4
national,	district,	project,	co-operator,

MISCELLANEOUS REMARKS DATA (4 types shown)

RECORD TYPE (C788) R M K S — RECORD SEQUENCE NO. (C311) — DATE OF REMARK (C184) month – day – year

REMARKS (C185)

Subsequent entries may be used to continue the remark. Miscellaneous remarks field is limited to 256 characters.

RECORD TYPE (C788) R M K S — RECORD SEQUENCE NO. (C311) — DATE OF REMARK (C184) month – day – year

REMARKS (C185)

Subsequent entries may be used to continue the remark. Miscellaneous remarks field is limited to 256 characters.

DISCHARGE DATA

RECORD SEQUENCE NO. (C147)

DATE DISCHARGE MEASURED (C148) month – day – year

TYPE OF DISCHARGE (C703) P pumped F flow

DISCHARGE (gpm) (C150)

ACCURACY OF DISCHARGE MEASUREMENT (C310)
E excellent (LT 2%), G good (2%-5%), F fair (5%-8%), P poor (GT 8%)

SOURCE OF DATA (C151)
A other gov't, D driller, G geologist, L logs, M memory, O owner, R other reported, S reporting agency, Z other

METHOD OF DISCHARGE MEASUREMENT (C152)
A acoustic meter, B bailer, C current meter, D Doppler meter, E estimated, F flume, M totaling meter, O orifice, P pitot-tube, R reported, T trajectory, U venturi meter, V volumetric meas, W weir, X unknown, Z other

PRODUCTION WATER LEVEL (C153)

STATIC WATER LEVEL (C154)

SOURCE OF DATA (C155)
A other gov't, D driller, G geologist, L logs, M memory, O owner, R other reported, S reporting agency, Z other

METHOD OF WATER-LEVEL MEASUREMENT (C156)
A airline, B recorder, C calibrated airline, D differential GP, E estimated, F transducer, G pressure gage, H calibrated press. gage, L geophysical logs, M manometer, N non-rec. gage, O observed, P acoustic pulse, R reported, S steel tape, T electric tape, V calibrated elec. tape, Z other

PUMPING PERIOD (C157)

SPECIFIC CAPACITY (C272)

DRAWDOWN (C309)

GEOHYDROLOGIC DATA

RECORD TYPE (C748) GEOH

RECORD SEQUENCE NO. (C721)

DEPTH TO TOP OF UNIT (C91)

DEPTH TO BOTTOM OF UNIT (C92)

UNIT IDENTIFIER (C93)

LITHOLOGY (C96)

CONTRIBUTING UNIT (C304)
P principal aquifer, Q aggregate of lithologic units, S secondary aquifer, N no contribution, U unknown

LITHOLOGIC MOD FIER (C97)

GEOHYDROLOGIC AQUIFER DATA

RECORD TYPE (C750) AQFR

RECORD SEQUENCE NO. (C742)

SEQUENCE NO. OF PARENT RECORD (C256)

DATE (C95) month – day – year

STATIC WATER LEVEL (C126)

CONTRIBUTION (C132)

SITE LOCATION SKETCH AND DIRECTIONS

Township ___________ Range ___________

Section # ___________

8 - Groundwater Site Schedule

GWPD 3—Establishing a permanent measuring point and other reference marks

VERSION: 2010.1

PURPOSE: To establish a permanent measuring point at a well from which water levels are measured, to establish a permanent land-surface datum, and to establish nearby reference marks.

Materials and Instruments

1. Groundwater Site Inventory (GWSI) System Groundwater Site Schedule, Form 9-1904-A
2. Measuring tape graduated in feet, tenths and hundredths of feet
3. Field notebook
4. Topographic map or Global Positioning System (GPS) receiver
5. Pencil or pen, blue or black ink. Strikethrough, date, and initial errors; no erasures
6. Spray paint, bright color or permanent marker
7. Metal file for marking well casing; hammer and cold steel chisel, survey monument (nail, spike, tablet)
8. Two wrenches with adjustable jaws or other tools for removing well cap
9. Key for well access
10. Camera
11. Protractor, calculator, or other tools to calculate angles and lengths
12. Rod, leveling instrument, and leveling notes sheets

Data Accuracy and Limitations

The "stickup" of a well is the length of well casing above the plane of the land-surface datum (LSD).

Altitude Accuracy: Vertical Stickup

The accuracy of the measuring point (MP) or LSD altitude depends on the measurement method used. When topographic maps are used, the accuracy typically is about one-half the contour interval of the topographic map. When geodetic differential GPS methods are used, the accuracy can be on the order of a couple of centimeters. When spirit leveling is used the accuracy is dependent on the order (1st, 2nd, 3rd) of surveying and the length of the survey line and typically can vary from tens of centimeters to a millimeter or less. Limitations: A high level of altitude accuracy is not critical when measurements obtained from a single well are compared to one another. Measurement accuracy is important, but altitude accuracy is not. If water-levels are to be compared *among wells*, however, a higher altitude accuracy (such as from spirit leveling) may be needed.

MP Correction Length Accuracy: Vertical Stickup

The MP correction length is the distance the measuring tape travels from the MP to the plane of the LSD (fig. 1). The accuracy of the MP correction length depends on the configuration of the MP with respect to the LSD. In the simplest example of a well with a vertical stickup and the LSD as a monument in the well pad or a file mark on the casing, the MP correction length can be measured directly with a measuring tape. In that instance, the accuracy of the measurement is 0.01 foot. In the case when the vertical distance between LSD and the MP cannot be directly measured with a tape, such as when a protective casing prevents direct measurement, the accuracy is a function of the measurement method used. A visual estimate using a measuring tape likely will have an accuracy slightly greater than 0.01 foot. When spirit leveling is used, the accuracy can vary from tens of centimeters to a millimeter or less. MP correction length accuracy is critical because a well may have more than one MP, all of which should be referenced to a single LSD. Limitations: Special considerations must be made for a well with a non-vertical stickup, when the configuration of the MP at the well does not allow the measuring tape to hang vertically directly from the MP through the plane of the LSD (fig. 2).

Altitude Accuracy: Non-Vertical Stickup

The altitude of the MP of a non-vertical stickup is not used directly, but may be measured for use in combination with the LSD altitude and the MP correction length. In the case of a non-vertical stickup, the accuracy of the LSD altitude is identical to that described in the vertical case. The accuracy of a water-level altitude calculated from the MP altitude and the MP correction length (option in Instruction no. 4) is equivalent to the least accurate measurement.

MP Correction Length Accuracy: Non-vertical Stickup

When the measurement tape does not hang vertically from the MP to the plane of the LSD, the MP correction length must be computed on the basis of the measurement path length and angles of deviation from vertical (fig. 2). The accuracy of this MP correction length is a function of the configuration of the well and the ability of the hydrographer to determine the tape path, but likely is greater than 0.01 foot.

Reference Mark Accuracy

A reference mark (RM) is used to determine whether the MP has moved with reference to LSD and, in extreme cases, to re-establish the LSD or MP at a well, thus the accuracy of the RM should be at least equivalent to that of the water-level measurement. In most instances, this is 0.01 foot. Limitation: comparability of water-level measurements made before and after re-establishment of the LSD or MP is limited by the accuracy of the RM.

Assumptions

1. For comparability to the water level measured in other wells, water-level measurements will be referenced consistently to the same vertical geodetic datum.
2. LSD is a specific type of RM. Once established, the LSD is not changed unless it is destroyed. If a new LSD must be established, the date of this change must be recorded, as well as the vertical distance between the destroyed LSD and the new LSD.
3. Measuring points change from time to time, especially on private wells. If a new MP must be established, the date of this change must be recorded, as well as the distance between the new MP and LSD (MP correction length).
4. Some wells have multiple measuring points or access points, especially production wells. Care must be taken in tracking these multiple MPs.
5. The operator can run leveling equipment in order to establish one or more RMs.

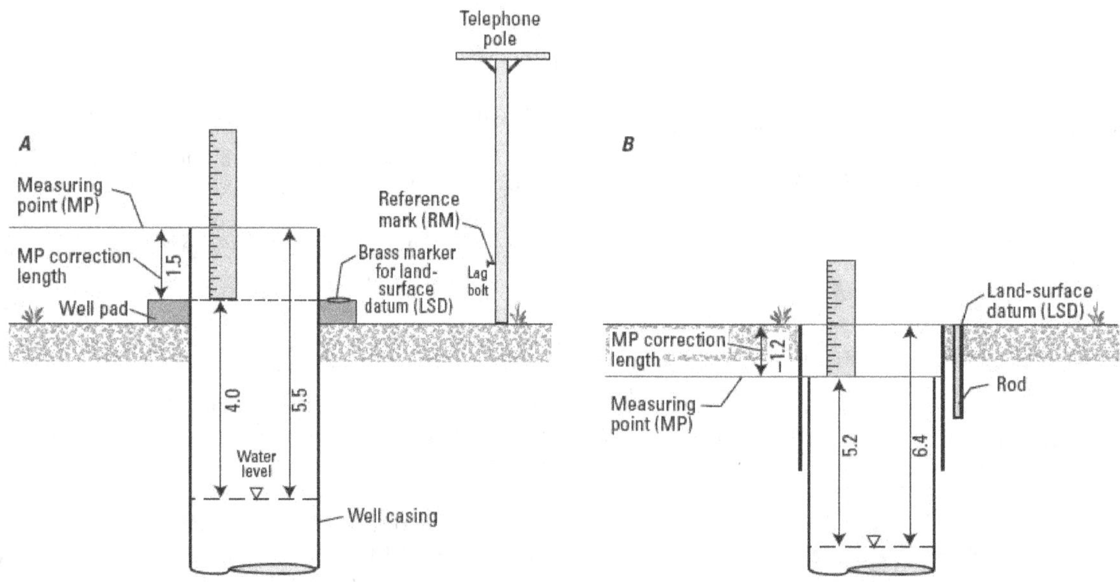

Figure 1. Relations among land-surface (LSD), measuring-point (MP), and reference-mark datums for measuring points above and below land surface. *A*, If the MP is above the LSD, subtract MP correction length to correct the water level to LSD (5.5 – 1.5 = 4.0). *B*, If the MP is below the LSD, subtract MP correction length to correct the water level to LSD (5.2 – (–1.2) = 6.4).

A

B

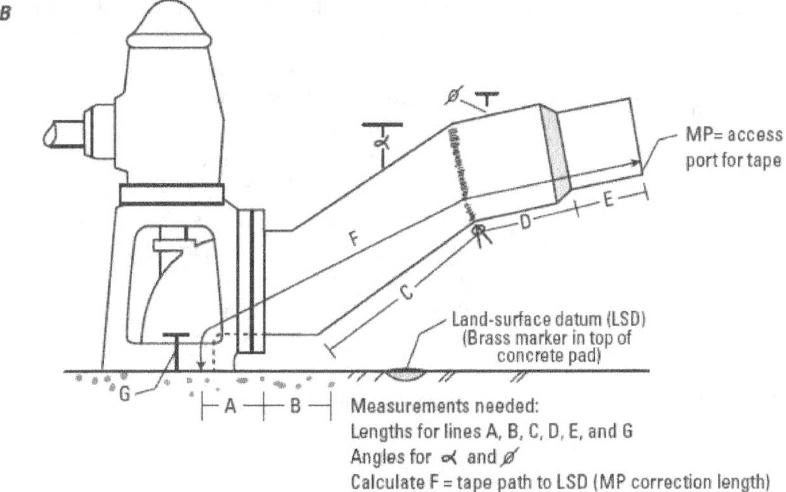

Figure 2. Examples of (*A*) determining a measuring point (MP) correction length when the configuration of the MP at the well does not allow the measuring tape to hang vertically directly from the MP through the plane of the land-surface datum (LSD) and (*B*) the measurements needed to calculate the MP correction length on the basis of the distance a tape would travel from the MP to the plane of the LSD in an irrigation well. (Photograph by E.L. Kuniansky, U.S. Geological Survey.)

Instructions

1. **Establish land-surface datum following these definitions and procedures:**

   a. The LSD at a well is a fixed RM at the well, at or near land surface, that can be used to measure the absolute vertical position (altitude) of the LSD and the distance from the LSD to the MP (the MP correction length).

   b. The LSD must be stable, as permanent as possible, clearly defined, clearly marked, and easily located.

   c. The LSD should be established to facilitate measuring from it to the MP.

   d. The LSD should be established to facilitate setting a survey rod or GPS antenna on the mark.

   e. Mark the LSD. For example, the LSD is noted by an 'X' etched into the well casing or is marked with a brass marker or chiseled "+" in the concrete pad at the base of the surface casing. If the landowner does not allow marking of the well, then describe the LSD as accurately as possible.

   f. Take a photograph of the LSD.

2. **Determine the altitude of the land-surface datum.**

    a. The altitude of the LSD must be determined for every site. At a minimum, it can be estimated from a topographic map. Locate the well using GWPD 5. Determine the altitude of the LSD from the topographic map.

    b. Optional: Depending on the use of the measurements from the well, the altitude of the LSD may be surveyed from a geodetic benchmark using spirit leveling or differential GPS techniques.

3. **Establish the measuring point following these definitions and procedures:**

    a. The MP is the most convenient place to measure the water level in a well. It is often at the top of the casing of an observation well, at the top of an access standpipe installed at a production well, or at an access point at the stem of a production well (see figs. 1 and 2).

    b. The MP must be stable, as permanent as possible, clearly defined, clearly marked, and easily located. For example, the MP is noted by a file mark on the well casing. The MP on a casing that does not have a horizontal rim commonly is established on the high or low side of the rim.

    c. If possible, position the MP at a particular point on the casing where a leveling rod could be set directly on it and the measuring tape can hang freely into the well when it is in contact with the MP.

    d. Using a file, lightly mark the MP on the well casing. Optionally, mark the MP by an arrow sprayed with a bright colored paint or permanent marker. If the MP cannot be marked, it must be clearly defined.

    e. Take a photograph of the MP.

    f. If more than one MP exists for a well, all MPs must be documented, and clearly differentiated.

    g. Optional: Depending on the use and storage of measurements from the well, the altitude of the MP of a well with a vertical stickup may be surveyed from a geodetic benchmark using spirit leveling or differential GPS techniques. MP altitude may be determined in two ways, depending on the calculation of the MP correction length described below.

4. **Determine the measuring point correction length following these definitions and procedures:**

    a. The MP correction length is the distance the measuring tape travels from the MP to the plane of the LSD. This is a vertical distance (also known as MP height) for a simple, vertical well. If the well stickup is not vertical, the MP correction length is not a true height above the LSD, but still represents the distance the tape must travel to reach the plane of the LSD.

    b. Measure the MP correction length in feet above or below the LSD (fig. 1). Values for MP correction lengths above LSD (fig. 1*A*) are positive numbers. Values for MP correction lengths below LSD (fig. 1*B*) are negative numbers and should be preceded by a minus sign (–).

        (1) For a well with a vertical stickup, where a water-level tape can hang vertically from the MP through the plane of the LSD (fig. 1), this distance can be measured directly with a steel tape or by leveling. Optional: if the objectives of the measurement require a precise altitude, the altitude of the MP for these wells can be surveyed from a geodetic benchmark using spirit leveling or differential GPS techniques.

        (2) For a well with a non-vertical stickup, where a water-level tape does not hang vertically from the MP through the plane of the LSD (fig. 2), the MP correction length cannot be measured directly. It is the distance between the MP and the plane of the LSD. The length along the measurement path between the MP and LSD must be computed on the basis of the measurement path length and angles of deviation from vertical (fig. 2). The geometry of this measurement path varies widely among this type of well. This will result in an MP correction length greater than the vertical distance between the LSD and the MP. Optional: If the objectives of the measurement require a precise water-level altitude, the altitude of the MP for wells with a non-vertical stickup should not be measured directly.

            (i) Water-level altitude can be referenced to the LSD, in which case the MP altitude is not needed.

            (ii) Water-level altitude can be referenced to the MP, in which case the MP altitude must be calculated by adding the MP correction length to the altitude of the LSD. Note that the MP altitude in this case is not a true altitude, but subtracting a depth to water measurement from this MP altitude will result in a true water-level altitude.

5. **Establish additional reference marks following these definitions and procedures:**

   a. An RM is a nearby datum established by permanent marks and is used to check the MP and (or) LSD or to re-establish the MP and (or) LSD should the original MP or LSD be destroyed or changed.

   b. Check the condition of the rod and leveling instrument.

   c. Establish the vertical relation between the MP and RMs by use of leveling (Kenney, 2010, for example). Establish at least one clearly marked RM near the well; more than one RM is preferable. For example, a benchmark, a lag bolt set in a telephone pole (fig. 1*A*), a spike in a mature tree, a mark on a permanent structure, or a poured concrete post. The RM should be located a suitable distance from the well to assure that a circumstance that damages a well does not also damage the RM.

   d. Take photographs of the RMs and include the photographs in the site field folder.

   e. A visual inspection of the MP, LSD, and RMs should be made at each site visit. Dates of any damage to the MP, LSD, or RMs must be documented. The vertical relation between the MP and RMs should be checked whenever there is evidence of damage to the MP, LSD, or RM. If no damage is apparent, the vertical relation between the MP and RMs should be confirmed at 3–5 year intervals.

Data Recording

Record data by use of appropriate field notebooks, level note sheets, and the GWSI Groundwater Site Schedule (fig. 3, Form 9-1904-A).

1. LSD: Record a description of the LSD in the field notebook, including the altitude, altitude accuracy, and geodetic datum. Final measurements should be documented in figure 3 as follows: (C16) Altitude of land surface, (C17) Method altitude determined, (C18) Altitude accuracy, and (C22) Altitude datum.

2. MP and MP correction length: Record a description of the MP in the field notebook, including the date of MP establishment, MP correction length or altitude, and a detailed description of the MP. Final data should be documented in figure 3 as follows: (C321) Beginning date, (C323) MP height (correction length), and (C324) MP remarks (description of the MP). If the altitude of the MP is determined, also record (C325) Measuring point altitude, (C326) Method altitude determined, (C327) Measuring point altitude accuracy, and (C328) Measuring point altitude datum. If an MP is destroyed or no longer in service, record the date of the destruction in (C322) Ending date.

3. RMs: Record a description of the site RMs in the field notebook, including the date of RM establishment. Document the vertical relation between the MP and RMs. Include the RM level notes in the site folder. Mark the MP and the RMs on the photographs and draw arrows to identify them. Store a copy of the photographs in the site folder.

References

Cunningham, W.L., and Schalk, C.W., comps., 2011a, Groundwater technical procedures of the U.S. Geological Survey, GWPD 1—Measuring water levels by use of a graduated steel tape: U.S. Geological Survey Techniques and Methods 1–A1, 4 p.

Cunningham, W.L., and Schalk, C.W., comps., 2011b, Groundwater technical procedures of the U.S. Geological Survey, GWPD 5—Documenting the location of a well: U.S. Geological Survey Techniques and Methods 1–A1, 10 p.

Hoopes, B.C., ed., 2004, User's manual for the National Water Information System of the U.S. Geological Survey, Ground-Water Site-Inventory System (version 4.4): U.S. Geological Survey Open-File Report 2005–1251, 274 p.

Kenney, T.A., 2010, Levels at gaging stations: U.S. Geological Survey Techniques and Methods 3–A19, 60 p.

U.S. Geological Survey, Office of Water Data Coordination, 1977, National handbook of recommended methods for water-data acquisition: Office of Water Data Coordination, Geological Survey, U.S. Department of the Interior, chap. 2, 149 p.

FORM NO. 9-1904-A
Revised Sept 2009, NWIS 4.9

File Code ______
Date ______

Coded by ______
Checked by ______
Entered by ______

U.S DEPT. OF THE INTERIOR
GEOLOGICAL SURVEY

GROUNDWATER SITE SCHEDULE
General Site Data

AGENCY CODE (C4) USGS
SITE ID (C1)
PROJECT (C5)

STATION NAME (C12/900)

[1] SITE TYPE (C802) Primary - Secondary
DISTRICT (C6)
COUNTRY (C41)
STATE (C7)
COUNTY or TOWN (C8) ______ County code

LATITUDE (C9) .
LONGITUDE (C10) .
LAT/LONG ACCURACY (C11): H (Hndrth sec.), 1 (tenth sec.), 5 (half sec.), S (sec.), R (3 sec.), F (6 sec.), T (10 sec.), M (min.), U (Un-known)

LAT/LONG METHOD (C35): C (land net), D (DGPS), G (GPS), L (LORAN), M (map), N (inter-polated digital map), R (reported), S (survey), U (un-known)
LAT/LONG DATUM (C36): NAD27 (North American Datum of 1927), NAD83 (North American Datum of 1983)
ALTITUDE (C16) .

ALTITUDE ACCURACY (C18)
ALTITUDE METHOD (C17): A (altimeter), D (DGPS), G (GPS), I (IfSAR), J (LIDAR), L (Level), M (map), N (DEM), R (re-ported), U (un-known)
ALTITUDE DATUM (C22): NGVD29 (National Geodetic Vertical Datum of 1929), NAVD88 (North American Vertical Datum of 1988)

LAND NET (C13): ¼ ¼ ¼ S section T township range merid

TOPO-GRAPHIC SETTING (C19): A (alluvial fan), B (playa), C (stream channel), D (depres-sion), E (dunes), F (flat), G (flood-plain), H (hill-top), K (sink-hole), L (lake or swamp), M (mangrove swamp), O (off-shore), P (pedi-ment), S (hill-side), T (ter-race), U (undu-lating), V (valley flat), W (upland draw)

HYDROLOGIC UNIT CODE (C20)
DRAINAGE BASIN CODE (C801)
STANDARD TIME ZONE (C813)
DAYLIGHT SAVINGS TIME FLAG (C814) Y OR **N**

MAP NAME (C14)
MAP SCALE (C15)

AGENCY USE (C803): A (active no/na), D (discon-tinued), I (inactive site), L (active written), M (active oral), O (inventory site), R (remediated)
[2] NATIONAL WATER-USE (C39)

DATA TYPE (C804) Place an 'A' (active), an 'I' (inactive), or an 'O' (inventory) in the appropriate box: WL cont, WL int, QW cont, QW int, PR cont, PR int, EV cont, EV int, wind vel., tide cont, tide int, sed. con, sed. ps, peak flow, low flow, state water use

INSTRUMENTS (C805) (Place a "Y" in the appropriate box): digital rec-order, graphic rec-order, tele-metry land line, tele-metry radio, tele-metry satellite, AHDAS, crest-stage gage, tide gage, deflec-tion meter, bubble gage, stilling well, CR type recorder, weigh-ing rain gage, tipping bucket rain gage, acoustic velocity meter, electro-magnetic flowmeter, pressure transducer

DATE INVENTORIED (C711) month – day – year
RECORD READY FOR WEB (C32): Y (ready to display), C (condi-tional), P (proprie-tary), L (local use only)

REMARKS (C806)

FOOTNOTES

[1] SITE TYPE (C802)

GL	Glacier	OC	Ocean	GW	Well	SB	Subsurface
WE	Wetland	OC-CO	Coastal	GW-CR	Collector or Ranney type well	SB-CV	Cave
AT	Atmosphere	LK	Lake, Reservoir, Impoundment	GW-EX	Extensometer well	SB-GWD	Groundwater drain
ES	Estuary			GW-HZ	Hyporheic -zone well	SB-TSM	Tunnel, shaft, or mine
LA	Land	SP	Spring	GW-IW	Interconnected wells	SB-UZ	Unsaturated zone
LA-EX	Excavation	ST	Stream	GW-TH	Test hole not completed as a well		
LA-OU	Outcrop	ST-CA	Canal	GW-MW	Multiple wells		
LA-SNK	Sinkhole	ST-DCH	Ditch				
LA-SH	Soil hole	ST-TS	Tidal stream				
LA-SR	Shore	FA-WIW	Waste-Injection well				

[2] WS (water supply), DO (domestic), CO (commer-cial), IN (industrial), IR (irrigation), MI (mining), LV (livestock), PH (power hydro-electric), ST (waste water treatment), RM (remedia-tion), TE (thermo-electric power), AQ (aqua-culture)

C22 Other (see manual for codes)
C36 Other (see manual for codes)
C39 is mandatory for all sites having data in SWUDS.

Figure 3. Groundwater Site Schedule, Form 9-1904-A.

GENERAL SITE DATA

DATA RELIABILITY (C3): C (field checked), L (poor location), M (minimal data), U (unchecked)

DATE OF FIRST CONSTRUCTION (C21): month – day – year

USE OF SITE (C23): A (anode), C (standby emer. supply), D (drain), E (geothermal), G (seismic), H (heat reservoir), M (mine), O (observation), P (oil or gas), R (recharge), S (repressurize), T (test), U (unused), V (withdrawal/return), W (withdrawal), X (waste), Z (destroyed)

SECONDARY USE OF SITE (C301) (See use of site)

TERTIARY USE OF SITE (C302) (See use of site)

USE OF WATER (C24): A (air cond.), B (bottling), C (commercial), D (dewater), E (power), F (fire), H (domestic), I (irrigation), J (industrial (cooling)), K (mining), M (medicinal), N (industrial), P (public supply), Q (aquaculture), R (recreations), S (stock), T (institutional), U (unused), Y (desalination), Z (other)

SECONDARY USE OF WATER (C25) (see use of water)

TERTIARY USE OF WATER (C26) (see use of water)

AQUIFER TYPE (C713): U (unconfined single), N (unconfined multiple), C (confined single), M (confined multiple), X (mixed)

PRIMARY AQUIFER (C714)

NATIONAL AQUIFER (C715)

HOLE DEPTH (C27)

WELL DEPTH (C28)

SOURCE OF DEPTH DATA (C29): A (other gov't), D (driller), G (geologist), L (logs), M (memory), O (owner), R (other reported), S (reporting agency), Z (other)

WATER-LEVEL DATA

DATE WATER-LEVEL MEASURED (C235): month – day – year

TIME (C709)

WATER-LEVEL TYPE CODE (C243): L (land surface), M (meas. pt.), S (vertical datum)

WATER LEVEL (C237/241/242)

MP SEQUENCE NO. (C248) (Mandatory if WL type=M)

WATER-LEVEL DATUM (C245) (Mandatory if WL type=S): NGVD29 (National Geodetic Vertical Datum Of 1929), NAVD88 (North American Vertical Datum Of 1988), Other (See manual for codes)

SITE STATUS FOR WATER LEVEL (C238): A (atmos. pressure), B (tide stage), C (ice), D (dry), E (recently flowing), F (flowing), G (nearby flowing), H (nearby recently flowing), I (injector site), J (injector site monitor), M (plugged), N (measurement discontinued), O (obstruction), P (pumping), R (recently pumped), S (nearby pumping), T (nearby recently pumped), V (foreign substance), W (well destroyed), X (affected by surface water), Z (other)

METHOD OF WATER-LEVEL MEASUREMENT (C239): A (airline), B (analog), C (calibrated airline), D (differential GPS), E (estimated), F (transducer), G (pressure gage), H (calibrated press. gage), L (geophysical logs), M (manometer), N (non-rec. gage), O (observed), P (acoustic pulse), R (reported), S (steel tape), T (electric tape), V (calibrated elec. tape), Z (other)

WATER-LEVEL ACCURACY (C276): 0 (foot), 1 (tenth), 2 (hundredth), 9 (not to nearest foot)

SOURCE OF WATER-LEVEL DATA (C244): A (other gov't), D (driller's log), G (geologist), L (geophysical logs), M (memory), O (owner), R (other reported), S (reporting agency), Z (other)

PERSON MAKING MEASUREMENT (C246) (WATER LEVEL PARTY)

MEASURING AGENCY (C247) (SOURCE)

EQUIP ID (C249) (20 char)

REMARKS (C267) (256 char)

RECORD READY FOR WEB (C858): Y (ready to display), C (conditional), P (proprietary), L (local use only)

CONSTRUCTION DATA

RECORD TYPE (C754): CONS

RECORD SEQUENCE NO. (C723)

DATE OF COMPLETED CONSTRUCTION (C60): month – day – year

NAME OF CONTRACTOR (C63)

SOURCE OF DATA (C64): A (other gov't), D (driller), G (geologist), L (logs), M (memory), O (owner), R (other reported), S (reporting agency), Z (other)

METHOD OF CONSTRUCTION (C65): A (air-rotary), B (bored or augered), C (cable tool), D (dug), H (hydraulic rotary), J (jetted), P (air percussion), R (reverse rotary), S (sonic), T (trenching), V (driven), W (drive wash), Z (other)

TYPE OF FINISH (C66): C (porous concrete), F (gravel w/perf.), G (gravel screen), H (horiz. gallery), O (open end), P (perf or slotted), S (screen), T (sand point), W (walled), X (open hole), Z (other)

TYPE OF SEAL (C67): B (bentonite), C (clay), G (cement grout), N (none), Z (other)

BOTTOM OF SEAL (C68)

METHOD OF DEVELOPMENT (C69): A (air-lift pump), B (bailed), C (compressed air), J (jetted), N (none), P (pumped), S (surged), Z (other)

HOURS OF DEVELOPMENT (C70)

SPECIAL TREATMENT (C71): C (chemicals), D (dry ice), E (explosives), F (deflocculent), H (hydrofracturing), M (mechanical), Z (other)

2 - Groundwater Site Schedule

CONSTRUCTION HOLE DATA (3 sets shown)

RECORD TYPE (C756) HOLE RECORD SEQUENCE NO. (C724) SEQUENCE NO. OF PARENT RECORD (C59)

DEPTH TO TOP OF INTERVAL (C73) . DEPTH TO BOTTOM OF NTERVAL (C74) . DIAMETER OF INTERVAL (C75) .

RECORD SEQUENCE NO. (C724)

DEPTH TO TOP OF INTERVAL (C73) . DEPTH TO BOTTOM OF NTERVAL (C74) . DIAMETER OF INTERVAL (C75) .

RECORD SEQUENCE NO. (C724)

DEPTH TO TOP OF INTERVAL (C73) . DEPTH TO BOTTOM OF NTERVAL (C74) . DIAMETER OF INTERVAL (C75) .

CONSTRUCTION CASING DATA (4 sets shown)

RECORD TYPE (C758) CSNG RECORD SEQUENCE NO. (C725) SEQUENCE NO. OF PARENT RECORD (C59)

DEPTH TO TOP OF CASING (C77) . DEPTH TO BOTTOM OF CASING (C78) . DIAMETER OF CASING (C79) .

[4] CASING MATERIAL (C80) CASING THICKNESS (C81) .

RECORD SEQUENCE NO. (C725) SEQUENCE NO. OF PARENT RECORD (C59)

DEPTH TO TOP OF CASING (C77) . DEPTH TO BOTTOM OF CASING (C78) . DIAMETER OF CASING (C79) .

[4] CASING MATERIAL (C80) CASING THICKNESS (C81) .

RECORD SEQUENCE NO. (C725) SEQUENCE NO. OF PARENT RECORD (C59)

DEPTH TO TOP OF CASING (C77) . DEPTH TO BOTTOM OF CASING (C78) . DIAMETER OF CASING (C79) .

[4] CAS NG MATERIAL (C80) CAS NG THICKNESS (C81) .

RECORD SEQUENCE NO. (C725) SEQUENCE NO. OF PARENT RECORD (C59)

DEPTH TO TOP OF CASING (C77) . DEPTH TO BOTTOM OF CASING (C78) . DIAMETER OF CASING (C79) .

[4] CAS NG MATERIAL (C80) CAS NG THICKNESS (C81) .

FOOTNOTE:

[4] CAS NG MATERIAL CODES

A	B	C	D	E	F	G	H	I	J	K	L	M	N	P	Q	R	S	T	U	V	W	X	Y	Z	4	6
abs	brick	concrete	copper	PTFE	Fiber-glass	galv. iron	Fiber-glass plastic	wrought iron	Fiber-glass epoxy	PVC thread-ed	glass	other metal	PVC glued	PVC or plastic	FEP	rock or stone	steel	tile	coated steel	stain-less steel	wood	steel carbon	steel galva-nized	other mat.	stain-less 304	stain-less 316

CONSTRUCTION OPENINGS DATA (3 sets shown)

RECORD TYPE (C760) OPEN   RECORD SEQUENCE NO. (C726)   SEQUENCE NO. OF PARENT RECORD (C59)

DEPTH TO TOP OF INTERVAL (C83)   DEPTH TO BOTTOM OF INTERVAL (C84)   DIAMETER OF INTERVAL (C87)

[5] MATERIAL TYPE (C86)   [6] TYPE OF OPENING (C85)   LENGTH OF OPENING (C89)   WIDTH OF OPENING (C88)

RECORD SEQUENCE NO. (C726)

DEPTH TO TOP OF INTERVAL (C83)   DEPTH TO BOTTOM OF INTERVAL (C84)   DIAMETER OF INTERVAL (C87)

[5] MATERIAL TYPE (C86)   [6] TYPE OF OPENING (C85)   LENGTH OF OPENING (C89)   WIDTH OF OPENING (C88)

RECORD SEQUENCE NO. (C726)

DEPTH TO TOP OF INTERVAL (C83)   DEPTH TO BOTTOM OF INTERVAL (C84)   DIAMETER OF INTERVAL (C87)

[5] MATERIAL TYPE (C86)   [6] TYPE OF OPENING (C85)   LENGTH OF OPENING (C89)   WIDTH OF OPENING (C88)

FOOTNOTES:

[5] TYPE OF MATERIAL CODES FOR OPEN SECTIONS

A	B	C	D	E	F	G	H	I	J	K	L	M	N	P	Q	R	S	T	V	W	X	Y	Z	4	6
ABS	brass or bronze	concrete	ceramic	PTFE	fiber-glass	galv. iron	fiber-glass plastic	wrought iron	fiber-glass epoxy	PVC thread-ed	glass	other metal	PVC glued	PVC	FEP	stain-less steel	steel	tile	brick	mem-brane	steel carbon	steel galva-nized	other	stain-less 304	stain-less 316

[6] TYPE OF OPENINGS CODES

F	L	M	P	R	S	T	W	X	Z
fractured rock	louvered or shutter-type	mesh screen	perforated, porous or slotted	wire-wound screen	screen (unk.)	sand point screen	walled or shored	open hole	other

CONSTRUCTION MEASURING POINT DATA

RECORD TYPE (C766) MPNT   RECORD SEQUENCE NO. (C728)   BEGINNING DATE (C321) month – day – year   ENDING DATE (C322)

M.P. HEIGHT (C323)   ALTITUDE OF MEASURING POINT (C325)   ALTITUDE METHOD (C326)   ALTITUDE ACCURACY (C327)

ALTITUDE DATUM (C328)   M.P. REMARKS (C324)

RECORD READY FOR WEB (C857)

Y	C	P	L
ready to display	condi-tional	proprie-tary	local use only

4 - Groundwater Site Schedule

CONSTRUCTION LIFT DATA

RECORD TYPE (C752) L I F T   RECORD SEQUENCE NO. (C254)   TYPE OF LIFT (C43) A air, B bucket, C centri-fugal, J jet, P piston, R rotary, S submer-sible, T turbine, U un-known, X no lift, Z other

DATE RECORDED (C38) month – day – year   PUMP INTAKE DEPTH (C44)   TYPE OF POWER (C45) D diesel, E electric, G gaso-line, H hand, L LP gas, N natural gas, S solar, W windmill, Z other

HORSE-POWER RATING (C46) .   MANUFACTURER (C48)   SERIAL NO. (C49)

POWER COMPANY (C50)   POWER COMPANY ACCOUNT NUMBER (C51)

POWER METER NUMBER (C52)   PUMP RATING (C53) (million gallons/units of fuel) .   ADDITIONAL LIFT (C255)

PERSON OR COMPANY MAINTAINING PUMP (C54)   RATED PUMP CAPACITY (gpm) (C268)   STANDBY POWER (C56) (see TYPE OF POWER)

HORSEPOWER OF STANDBY POWER SOURCE (C57) .

MISCELLANEOUS OWNER DATA

RECORD TYPE (C768) O W N R   RECORD SEQUENCE NO. (C718)   DATE OF OWNERSHIP (C159) – –

WU OWNER TYPE (C350) CP Corporation, GV Govern-ment, IN Individual, MI Military, OT Other, TG Tribal, WS Water Supplier   END DATE OF OWNERSHIP (C374) – –

OWNER'S NAME (C161)

EXAMPLES: JONES, RALPH A.
JONES CONSTRUCTION COMPANY

OWNER'S PHONE NUMBER (C351)   ACCESS TO OWNER'S NAME (C352) 0 Public Access, 1 Coop-erator, 2 USGS Only, 3 District Only, 4 Proprietary

OWNER'S ADDRESS (LINE 1) (C353)

OWNER'S ADDRESS (LINE 2) (C354)

OWNER'S CITY NAME (C355)

STATE (C356)   OWNER'S ZIP CODE (C357) –

OWNER'S COUNTRY NAME (C358)

ACCESS TO OWNER'S PHONE/ADDRESS (C359) 0 Public Access, 1 Coop-erator, 2 USGS Only, 3 District Only, 4 Proprietary

MISCELLANEOUS VISIT DATA

RECORD TYPE (C774) V I S T   RECORD SEQUENCE NO. (C737)   DATE OF VISIT (C187) month – day – year

NAME OF PERSON (C188)

MISCELLANEOUS OTHER ID DATA (2 sets shown)

RECORD TYPE (C770) O T I D

RECORD SEQUENCE NO. (C736)

OTHER ID (C190)

ASSIGNER (C191)

RECORD SEQUENCE NO. (C736)

OTHER ID (C190)

ASSIGNER (C191)

MISCELLANEOUS OTHER DATA

RECORD TYPE (C772) O T D T

RECORD SEQUENCE NO. (C312)

OTHER DATA TYPE (C181)

OTHER DATA LOCATION (C182) C Cooperator's Office, D District Office R Reporting Agency Z other

DATA FORMAT (C261) F files, M machine readable, P published, Z other

MISCELLANEOUS LOGS DATA (3 sets shown)

RECORD TYPE (C778) L O G S

RECORD SEQUENCE NO. (C739)

TYPE OF LOG (C199)

BEG NN NG DEPTH (C200) .

ENDING DEPTH (C201) .

SOURCE OF DATA (C202) A other gov't D driller G geol-ogist L logs M memory O owner R other reported S reporting agency Z other

DATA FORMAT (C225) F files M machine readable P published Z other

OTHER DATA LOCATION (C226)

RECORD TYPE (C778) L O G S

RECORD SEQUENCE NO. (C739)

TYPE OF LOG (C199)

BEG NNING DEPTH (C200) .

ENDING DEPTH (C201) .

SOURCE OF DATA (C202) A other gov't D driller G geol-ogist L logs M memory O owner R other reported S reporting agency Z other

DATA FORMAT (C225) F files M machine readable P published Z other

OTHER DATA LOCATION (C226)

RECORD TYPE (C778) L O G S

RECORD SEQUENCE NO. (C739)

TYPE OF LOG (C199)

BEGINNING DEPTH (C200) .

ENDING DEPTH (C201) .

SOURCE OF DATA (C202) A other gov't D driller G geol-ogist L logs M memory O owner R other reported S reporting agency Z other

DATA FORMAT (C225) F files M machine readable P published Z other

OTHER DATA LOCATION (C226)

ACOUSTIC LOG:
AS Sonic
AV Acoustic velocity
AW Acoustic waveform
AT Acous ic televiewer

CALIPER LOG:
CP Caliper
CS Caliper, single arm
CT Caliper, three arm
CM Caliper, multi arm
CA Caliper, acous ic

DRILLING LOG:
DT Drilling ime
DR Drillers
DG Geologists
DC Core

ELECTRIC LOG:
EE Electric
ER Single-point resistance
EP Spontaneous potential
EL Long-normal resis ivity
ES Short-normal resis ivity
EF Focused resistivity
ET Lateral resistivity
EN Microresistivity
EC Microresistivity, forused
EO Microresis ivity, lateral
ED Dipmeter

ELECTROMAGNETIC LOG:
MM Magnetic log
MS Magnetic susceptibiity log
MI Electromagne ic induc ion log
MD Electromagne ic dual induction log
MR Radar reflection image log
MV Radar direct-wave velocity log
MA Radar direct-wave amplitude log

FLUID LOG:
FC Fluid conductivity
FR Fluid resis ivity
FT Fluid temperature
FF Fluid differential temperature
FV Fluid velocity
FS Spinner flowmeter
FH Heat-pulse flowmeter
FE Electromagnetic flowmeter
FD Doppler flowmeter
FA Radioactive tracer
FY Dye tracer
FB Brine tracer

NUCLEAR LOG:
NG Gamma
NS Spectral gamma
NA Gamma-gamma
NN Neutron
NT Neutron activitation
NM Neuclear magnetic resonance

OPTICAL LOG:
OV Video
OF Fisheye video
OS Sidewall video
OT Optical televiewer

COMBINATION LOG:
ZF Gamma, fluid resistivity, temperature
ZI Gamma, electromagne ic induction
ZR Long/short normal resistivity
ZT Fluid resistivity, temperature
ZM Electromagnetic flowmeter, fluid resistivity, temperature
ZN Long/short normal resistivity, spontaneous poten ial
ZP Single-point resistance, spontaneous potential
ZE Gamma, long/short normal resistivity, spontaneous potential, single-point resistance, fluid resi ivity, temperature

WELL CONSTRUCTION LOG:
WC Casing collar
WD Borehold deviation

OTHER LOG:
OR Other

6 - Groundwater Site Schedule

MISCELLANEOUS NETWORK DATA (3 types shown)

RECORD TYPE (C780) N E T W | RECORD SEQUENCE NO. (C730) | TYPE OF NETWORK (C706) Q W water quality | BEGINNING YEAR (C115) | ENDING YEAR (C116)

TYPE OF ANALYSIS (C120)

A	B	C	D	E	F	G	H	I	J	K	L	M	N	P	Z
physical proper-ties	common ions	trace elements	pesti-cides	nutri-ents	sanitary analysis	codes D&B	codes B&E	codes B&C	codes B&F	codes D&E	codes C,D&E	all or most	codes B&C& radio-active	codes B,C&A	other

SOURCE AGENCY (C117) | [7] FREQUENCY OF COLLECTION (C118) | ANALYZING AGENCY (C307) | [8] PRIMARY NETWORK SITE (C257) | [8] SECONDARY NETWORK SITE (C708)

RECORD TYPE (C780) N E T W | RECORD SEQUENCE NO. (C730) | TYPE OF NETWORK (C706) W L water level | BEGINNING YEAR (C115) | ENDING YEAR (C116)

SOURCE AGENCY (C117) | [7] FREQUENCY OF COLLECTION (C118) | [8] PRIMARY NETWORK SITE (C257) | [8] SECONDARY NETWORK SITE (C708)

RECORD TYPE (C780) N E T W | RECORD SEQUENCE NO. (C730) | TYPE OF NETWORK (C706) W D pumpage or with-drawals | BEGINNING YEAR (C115) | ENDING YEAR (C116)

SOURCE AGENCY (C117) | [7] FREQUENCY OF COLLECTION (C118) | METHOD OF COLLECTION (C133) | [8] PRIMARY NETWORK SITE (C257) | [8] SECONDARY NETWORK SITE (C708)

C	E	M	U	Z
calcu-lated	esti-mated	meter-ed	un-known	other

FOOTNOTES:

[7] FREQUENCY OF COLLECTION CODES

A	B	C	D	F	I	M	O	Q	S	W	Z	2	3	4	5	X
annually	bi monthly	continu-ously	daily	semi-monthly	inter mittent	monthly	one-time only	quarter-ly	semi-annually	weekly	other	bi-annually	every 3 years	every 4 years	every 5 years	every 10 years

[8] NETWORK SITE CODES

1	2	3	4
national,	district,	project,	co-operator,

MISCELLANEOUS REMARKS DATA (4 types shown)

RECORD TYPE (C788) R M K S | RECORD SEQUENCE NO. (C311) | DATE OF REMARK (C184) month – day – year

REMARKS (C185)

Subsequent entries may be used to continue the remark. Miscellaneous remarks field is limited to 256 characters.

RECORD TYPE (C788) R M K S | RECORD SEQUENCE NO. (C311) | DATE OF REMARK (C184) month – day – year

REMARKS (C185)

Subsequent entries may be used to continue the remark. Miscellaneous remarks field is limited to 256 characters.

DISCHARGE DATA

RECORD SEQUENCE NO. (C147)

DATE DISCHARGE MEASURED (C148) month – day – year

TYPE OF DISCHARGE (C703) P pumped F flow

DISCHARGE (gpm) (C150)

ACCURACY OF DISCHARGE MEASUREMENT (C310): E excellent (LT 2%), G good (2%-5%), F fair (5%-8%), P poor (GT 8%)

SOURCE OF DATA (C151): A other gov't, D driller, G geologist, L logs, M memory, O owner, R other reported, S reporting agency, Z other

METHOD OF DISCHARGE MEASUREMENT (C152): A acoustic meter, B bailer, C current meter, D Doppler meter, E estimated, F flume, M totaling meter, O orifice, P pitot-tube, R reported, T trajectory, U venturi meter, V volumetric meas, W weir, X unknown, Z other

PRODUCTION WATER LEVEL (C153)

STATIC WATER LEVEL (C154)

SOURCE OF DATA (C155): A other gov't, D driller, G geologist, L logs, M memory, O owner, R other reported, S reporting agency, Z other

METHOD OF WATER-LEVEL MEASUREMENT (C156): A airline, B recorder, C calibrated airline, D differential GP, E estimated, F transducer, G pressure gage, H calibrated press. gage, L geophysical logs, M manometer, N non-rec. gage, O observed, P acoustic pulse, R reported, S steel tape, T electric tape, V calibrated elec. tape, Z other

PUMP NG PERIOD (C157)

SPECIFIC CAPACITY (C272)

DRAWDOWN (C309)

GEOHYDROLOGIC DATA

RECORD TYPE (C748) G E O H

RECORD SEQUENCE NO. (C721)

DEPTH TO TOP OF UNIT (C91)

DEPTH TO BOTTOM OF UNIT (C92)

UNIT IDENT FIER (C93)

LITHOLOGY (C96)

CONTRIBUTING UNIT (C304): P principal aquifer, Q aggregate of lithologic units, S secondary aquifer, N no contribution, U unknown

LITHOLOGIC MOD FIER (C97)

GEOHYDROLOGIC AQUIFER DATA

RECORD TYPE (C750) A Q F R

RECORD SEQUENCE NO. (C742)

SEQUENCE NO. OF PARENT RECORD (C256)

DATE (C95) month – day – year

STATIC WATER LEVEL (C126)

CONTR BUTION (C132)

SITE LOCATION SKETCH AND DIRECTIONS

Township ___________ Range ___________

Section #___________

8 - Groundwater Site Schedule

GWPD 4—Measuring water levels by use of an electric tape

VERSION: 2010.1

PURPOSE: To measure the depth to the water surface below land-surface datum using the electric tape method.

Materials and Instruments

1. An electric tape, double-wired and graduated in feet, tenths and hundredths of feet. Electric tapes commonly are mounted on a hand-cranked and powered supply reel that contains space for the batteries and some device ("indicator") for signaling when the circuit is closed (fig. 1).
2. An older model electric tape, also known as an "M-scope," marked at 5-foot intervals with clamped-on metal bands (fig. 2) has been replaced by newer, more accurate models. Technical procedures for this device are available from the procedures document archives.
3. A steel reference tape for calibration, graduated in feet, tenths and hundredths of feet
4. Electric tape calibration and maintenance equipment logbook
5. Pencil or pen, blue or black ink. Strikethrough, date, and initial errors; no erasures
6. Water-level measurement field form, or handheld computer for data entry
7. Two wrenches with adjustable jaws or other tools for removing well cap
8. Key for well access
9. Clean rag
10. Cleaning supplies for water-level tapes as described in the National Field Manual (Wilde, 2004)
11. Replacement batteries

Data Accuracy and Limitations

1. A modern graduated electric tape commonly is accurate to +/– 0.01 foot.
2. Most accurate for water levels less than 200 feet below land surface.
3. The electric tape should be calibrated against an acceptable steel tape. An acceptable steel tape is one that is maintained in the office for use only for calibrating tapes, and this calibration tape never is used in the field.
4. If the water in the well has very low specific conductance, an electric tape may not give an accurate reading.
5. Material on the water surface, such as oil, ice, or debris, may interfere with obtaining consistent readings.
6. Corrections are necessary for measurements made from angled well casings.
7. When measuring deep water levels, tape expansion and stretch is an additional consideration (Garber and Koopman, 1968).

Advantages

1. Superior to a steel tape when water is dripping into the well or condensing on the inside casing walls.
2. Superior to a steel tape in wells that are being pumped, particularly with large-discharge pumps, where the splashing of the water surface makes consistent results by the wetted-tape method impossible. Also safer to use in pumped wells because the water is sensed as soon as

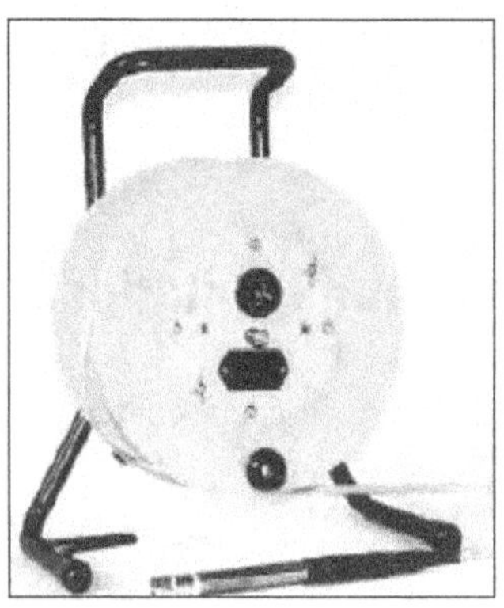

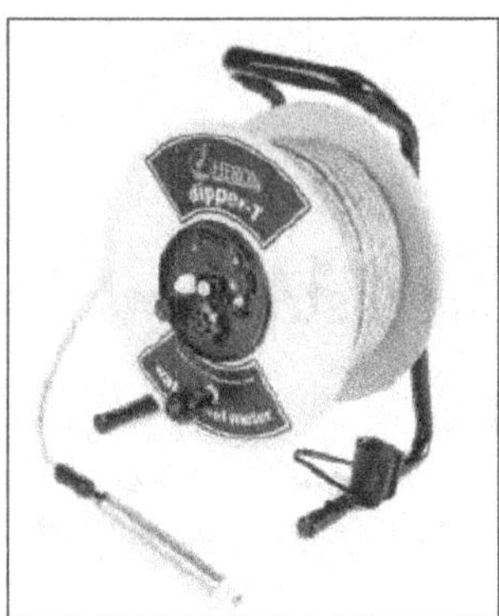

Figure 1. An electric tape or cable, double wired and marked the entire length in feet, tenths and hundredths of feet, that can be considered accurate to 0.01 foot at depths of less than 200 feet. Electric tapes commonly are mounted on a hand-cranked and powered supply real that contains space for the batteries and some device ("indicator") for signaling when the circuit is closed. Brand names are for illustration purposes only and do not imply endorsement by the U.S. Geological Survey. (Photographs used with permission of vendors.)

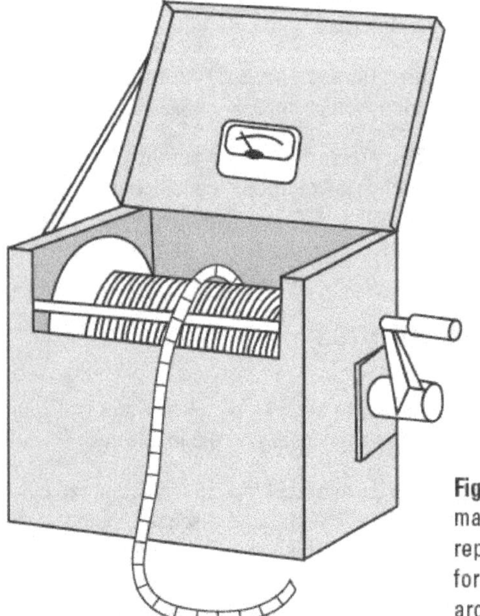

Figure 2. Older model electric tape, also known as "M-scope" marked at 5-foot intervals with clamped-on metal bands, has been replaced by newer, more accurate models. Technical procedures for this device are available from the procedures document archives.

the probe reaches the water surface and there is less danger of lowering the tape into the pump impellers.

3. Superior to a steel tape when a series of measurements are needed in quick succession, such as in aquifer tests, because the electric tape does not have to be removed from the well for each reading.

Disadvantages

1. Harder to keep calibrated than a steel tape.
2. Electric connections require maintenance.
3. Requires battery power.
4. Cable jacket is subject to wear and tear. Continuity of the electrical circuit must be maintained.

Assumptions

1. An established measuring point (MP) exists and the distance from the MP to the land-surface datum (LSD) is known. See GWPD 3 for the technical procedures on establishing a permanent MP.
2. The MP is clearly marked and described so that a person who has not measured the well will be able to recognize it.
3. The well is free of obstructions that could affect the plumbness of the steel tape and cause errors in the measurement.
4. The same field method is used for measuring depth below the MP, or depth relative to vertical datum, but with a different datum correction.
5. The tape is calibrated against a steel reference tape.
6. Field measurements will be recorded on paper forms. When using a handheld computer to record field measurements, the measurement procedure is the same, but the instructions below refer to a specific paper field form.

Tape Calibration And Maintenance

Before using an electric tape in the field, calibrate it against a steel reference tape. A reference tape is one that is maintained in the office only to calibrate other tapes.

1. Calibration of electric tape:
   - Check the distance from the probe's sensor to the nearest foot marker on the tape to ensure that this distance puts the sensor at the zero-foot point for the tape. If it does not, a correction must be applied to all depth-to-water measurements.
   - Compare length marks on the electric tape with those on the steel reference tape while the tapes are laid out straight on level ground, or compare the electric tape with a known distance between fixed points on level ground.
   - Compare water-level measurements made with the electric tape with those made with a calibrated steel tape in several wells that span the range of depths to water that is anticipated. Measurements should agree to within +/– 0.02 foot. If measurements are not repeatable to this standard, then a correction factor based on a regression analysis should be developed and applied to measurements made with the electric tape.
2. Using a repaired/spliced tape: If the tape has been repaired by cutting off a section of tape that was defective and splicing the sensor to the remaining section of the tape, then the depth to water reading at the MP will not be correct. To obtain the correct depth to water, apply the following steps, which is similar to the procedure for using a steel tape and chalk. Using the water-level measurement field form (fig. 3) to record these modifications:
   - Ensure that the splice is completely insulated from any moisture and that the electrical connection is complete.
   - Measure the distance from the sensing point on the probe to the nearest foot marker above the spliced section of tape. Subtract that distance from the nearest foot marker above the spliced section of tape. That value then becomes the "tape correction." For example, if the nearest foot marker above the splice is 20 feet, and the distance from that foot marker to the probe sensor is 0.85 foot, then the tape correction will be 19.15 feet. Write down the tape correction on the water-level measurement field form (fig. 3). Periodically recheck this value by measuring with the steel reference tape.
3. Maintain the tape in good working condition by periodically checking the tape for breaks, kinks, and possible stretch.
4. Carry extra batteries, and check battery strength regularly.
5. The electric tape should be recalibrated annually or more frequently if it is used often or if the tape has been subjected to abnormal stress that may have caused it to stretch.

WATER-LEVEL MEASUREMENT FIELD FORM

Calibrated Electric Tape Measurement

SITE INFORMATION

SITE ID (C1)

Equipment ID

Date of Field Visit

Station name (C12)

WATER-LEVEL DATA

	1	2	3	4	5
Time					
Hold					
Tape correction					
WL below MP					
MP correction					
WL below LSD					

Measured by ______________________ COMMENTS* ______________________

*Comments should include quality concerns and changes in: M.P., ownership, access, locks, dogs, measuring problems, et al.

MEASURING POINT DATA (for MP Changes)

M.P. REMARKS (C324)

BEGINNING DATE (C321)

month day year

ENDING DATE (C322)

M.P. HEIGHT (C323) NOTE: (-) for MP below land surface

Final Measurement for GWSI

WATER LEVEL TYPE CODE (C243) L M S

below land surface, below meas. pt., sea level

DATE WATER LEVEL MEASURED (C235) — month, day, year

TIME (C709)

STATUS (C238)

METHOD (C239)

TYPE (C243)

WATER LEVEL (C237)

METHOD OF WATER-LEVEL MEASUREMENT(C239)

A	B	C	E	G	H	L	M	N	R	S (GWPD1)	T	V (GWPD4)	Z
airline,	analog,	calibrated airline,	estimated,	pressure gage,	calibrated press. gage,	geophysical logs,	manometer,	non-rec. gage,	reported,	steel tape,	electric tape,	calibrated elec. tape	other

SITE STATUS FOR WATER LEVEL (C238)

D	E	F	G	H	I	J	M	N	O	P	R	S	T	V	W	X	Z	BLANK
dry,	recently flowing,	flowing,	nearby flowing	nearby recently flowing,	injector site,	injector site monitor,	plugged,	measurement discon.,	obstruction,	pumping,	recently pumped,	nearby pumping,	nearby recently pumped,	foreign sub-stance,	well des-troyed,	surface water effects,	other	static

Figure 3. Water-level measurement field form for calibrated electric tape measurements. This form, or an equivalent custom-designed form, should be used to record field measurements.

Instructions

1. Check the circuitry of the electric tape before lowering the probe into the well by dipping the probe into tap water and observing whether the indicator needle, light, and (or) beeper (collectively termed the "indicator" in this document) are functioning properly to indicate a closed circuit. If the tape has multiple indicators (sound and light, for instance), confirm that they are operating simultaneously. If they are not, determine the most accurate indicator.

2. Make all readings using the same deflection point on the indicator scale, light intensity, or sound so that water levels will be consistent among measurements.

3. Lower the electrode probe slowly into the well until the indicator shows that the circuit is closed and contact with the water surface is made (fig. 4). Place the nail of the index finger on the insulated wire at the MP and read the depth to water.

4. Record the date and time of the measurement. Record the depth to water measurement in the row "Hold" (fig. 3). If the tape has been repaired and spliced or has a calibration correction (see the section above on using a repaired/spliced tape), subtract the "Tape Correction" value from the "Hold" value, and record this difference in the row "WL below MP" (fig. 3).

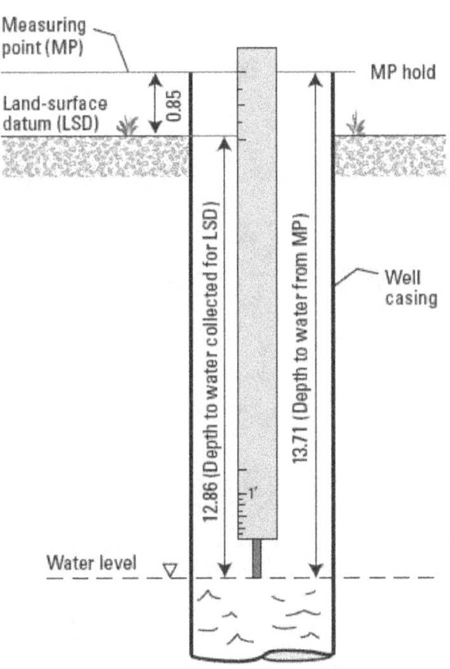

Figure 4. Water-level measurement using a graduated electric tape.

5. Record the MP correction length on the "MP correction" row of the field form (fig. 3). Subtract the MP correction length from the true "WL below MP" value to get the depth to water below or above LSD. The MP correction is positive if the MP is above land surface and is negative if the MP is below land surface (GWPD 3). Record the water level in the "WL below LSD" column of the water-level measurement field form (fig. 3). If the water level is above LSD, record the depth to water in feet above land surface as a negative number.

6. Pull the tape up and make a check measurement by repeating steps 3–5. Record the check measurement in column 2 of the field form. If the check measurement does not agree with the original measurement within 0.02 foot, continue to make measurements until the reason for lack of agreement is determined or the results are shown to be reliable. If more than two measurements are made, use best judgment to select the measurement most representative of field conditions. Complete the "Final Measurement for GWSI" portion of the field form.

7. After completing the water-level measurement, disinfect and rinse that part of the tape that was submerged below the water surface as described in the National Field Manual (Wilde, 2004). This will reduce the possibility of contamination of other wells from the tape. Rinse the tape thoroughly with deionized or tap water to prevent tape damage. Dry the tape and rewind onto the tape reel.

Data Recording

All calibration and maintenance data associated with the electric tape being used are recorded in the calibration and maintenance equipment logbook. All data are recorded in the water-level measurement field form (fig. 3) to the appropriate accuracy for the depth being measured.

References

Cunningham, W.L., and Schalk, C.W., comps., 2011a, Groundwater technical procedures of the U.S. Geological Survey, GWPD 1—Measuring water levels by use of a graduated steel tape: U.S. Geological Survey Techniques and Methods 1–A1, 4 p.

Cunningham, W.L., and Schalk, C.W., comps., 2011b, Groundwater technical procedures of the U.S. Geological Survey, GWPD 3—Establishing a permanent measuring point and other reference marks: U.S. Geological Techniques and Methods 1–A1, 13 p.

Garber, M.S., and Koopman, F.C., 1968, Methods of measuring water levels in deep wells: U.S. Geological Survey Techniques of Water-Resources Investigations, book 8, chap. A1, p. 6–11.

Heath, R.C., 1983, Basic ground-water hydrology: U.S. Geological Survey Water-Supply Paper 2220, p. 72–73.

Hoopes, B.C., ed., 2004, User's manual for the National Water Information System of the U.S. Geological Survey, Ground-Water Site-Inventory System (version 4.4): U.S. Geological Survey Open-File Report 2005–1251, 274 p.

U.S. Geological Survey, Office of Water Data Coordination, 1977, National handbook of recommended methods for water-data acquisition: Office of Water Data Coordination, Geological Survey, U.S. Department of the Interior, chap. 2, 149 p.

Wilde, F.D., ed., 2004, Cleaning of equipment for water sampling (version 2.0): U.S. Geological Survey Techniques of Water-Resources Investigations, book 9, chap. A3, accessed July 17, 2006, at *http://pubs.water.usgs.gov/twri9A3/*.

GWPD 5—Documenting the location of a well

VERSION: 2010.1

PURPOSE: To specify a procedure for documenting the location of a well at a groundwater site.

Materials and Instruments

1. Global Positioning System (GPS) receiver, if available
2. GPS calibration and maintenance equipment logbook
3. Best available paper maps:
    - A state highway map
    - Town or county plat map
    - An aerial photograph or satellite image
    - U.S. Geological Survey (USGS) 7.5-minute topographic quadrangle map
        - USGS 7.5-minute latitude-longitude scale
        - USGS 1:24,000 scale, graduated in miles and feet
4. Orienteering (transparent base) compass
5. Groundwater Site Inventory (GWSI) System Groundwater Site Schedule, Form 9-1904-A
6. Field notebook
7. Pencil or pen, blue or black ink. Strikethrough, date, and initial errors; no erasures
8. Camera

Data Accuracy and Limitations

1. GPS instrument accuracy varies. Handheld, Wide Area Augmentation System (WAAS)-enabled GPS instruments typically are accurate within a few meters horizontally. Instrument manuals and field tests should be used to confirm instrument accuracy.
2. USGS 7.5-minute latitude-longitude scale should be accurate to 0.5 second or about 50 feet.

Assumptions

1. The person locating the well has been trained to use a GPS instrument to determine the latitude and longitude of a point on the ground.
2. The person locating the well has been trained to use a latitude-longitude scale to determine the latitude and longitude of a point on a USGS 7.5-minute topographic quadrangle map.

Instructions

1. Each groundwater site should have a station log containing detailed narrative descriptions of the site, permanent landmarks, the best route to the site, and job hazards in the vicinity of the site.
2. Make two sketch maps of the site, one showing the general location of the site, and the other showing the details of the site. Orient the sketch maps relative to north using a compass. All distances should be shown in feet from permanent landmarks, such as buildings, bridges, culverts, telephone poles, road centerlines, and road intersections (fig. 1).
    a. General location map:
        (1) If a GPS instrument is available, determine the latitude and longitude of the well site.
        (2) Plot the general location of the well on a suitable paper map. If a GPS instrument is not available, the location should be plotted on a USGS 7.5-minute topographic quadrangle map.
        (3) If a GPS instrument is not available, determine the latitude and longitude of the well site from a USGS 7.5-minute topographic quadrangle map using a USGS 7.5-minute latitude-longitude scale.

A

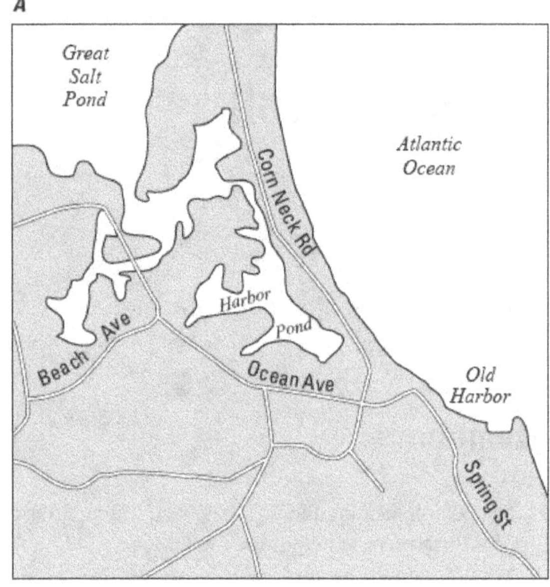

B

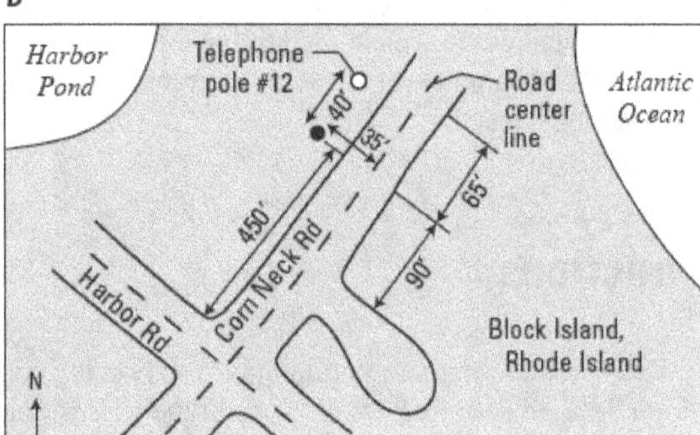

Figure 1. Examples of (*A*) general sketch map and (*B*) detailed sketch map.

b. Detailed site map:

(1) Prepare a detailed sketch map (fig. 1) showing the location of the well site in the field notebook and on the last page of the Groundwater Site Schedule, Form 9-1904-A (fig. 2). The sketch map should contain enough detail so that the site could be found by a person who has never been to the site before.

(2) Take at least two photographs of the well location from different views and indicate on each photograph the direction of view. File location photographs with the GWSI form.

Data Recording

All calibration and maintenance data associated with the GPS instrument use are recorded in the calibration and maintenance equipment logbook. Data are recorded in a field notebook and on the GWSI Groundwater Site Schedule (Form 9-1904-A).

References

Hoopes, B.C., ed., 2004, User's manual for the National Water Information System of the U.S. Geological Survey, Ground-Water Site-Inventory System (version 4.4): U.S. Geological Survey Open-File Report 2005–1251, 274 p.

U.S. Geological Survey, Office of Water Data Coordination, 1977, National handbook of recommended methods for water-data acquisition: Office of Water Data Coordination, Geological Survey, U.S. Department of the Interior, chap. 2, 149 p.

FORM NO. 9-1904-A
Revised Sept 2009, NWIS 4.9

File Code ____________
Date ____________

Coded by ____________
Checked by ____________
Entered by ____________

U.S DEPT. OF THE INTERIOR
GEOLOGICAL SURVEY

GROUNDWATER SITE SCHEDULE
General Site Data

AGENCY CODE (C4) USGS | SITE ID (C1) | PROJECT (C5)

STATION NAME (C12/900)

SITE TYPE (C802)[1] Primary - Secondary | DISTRICT (C6) | COUNTRY (C41) | STATE (C7)

COUNTY or TOWN (C8) ____________ County code

LATITUDE (C9) | LONGITUDE (C10) | LAT/LONG ACCURACY (C11): H Hndrth sec. 1 tenth sec. 5 half sec. S sec. R 3 sec. F 5 sec. T 10 sec. M min. U Un-known

LAT/LONG METHOD (C35): C land net, D DGPS, G GPS, L LORAN, M map, N inter-polated digital map, R reported, S survey, U un-known

LAT/LONG DATUM (C36): NAD27 North American Datum of 1927; NAD83 North American Datum of 1983

ALTITUDE (C16)

ALTITUDE ACCURACY (C18)

ALTITUDE METHOD (C17): A altimeter, D DGPS, G GPS, I IfSAR, J LIDAR, L Level, M map, N DEM, R re-ported, U un-known

ALTITUDE DATUM (C22): NGVD29 National Geodetic Vertical Datum of 1929; NAVD88 North American Vertical Datum of 1988

LAND NET (C13): ¼ ¼ ¼ S section T township range merid

TOPOGRAPHIC SETTING (C19): A alluvial fan, B playa, C stream channel, D depres-sion, E dunes, F flat, G flood-plain, H hill-top, K sink-hole, L lake or swamp, M mangrove swamp, O off-shore, P pedi-ment, S hill-side, T ter-race, U undu-lating, V valley flat, W upland draw

HYDROLOGIC UNIT CODE (C20) | DRAINAGE BASIN CODE (C801) | STANDARD TIME ZONE (C813) | DAYLIGHT SAVINGS TIME FLAG (C814) Y OR **N**

MAP NAME (C14) | MAP SCALE (C15)

AGENCY USE (C803): A active no/na, D discon-tinued, I inactive site, L active written, M active oral, O inventory site, R remediated

[2] NATIONAL WATER-USE (C39)

DATA TYPE (C804)
Place an 'A' (active), an 'I' (inactive), or an 'O' (inventory) in the appropriate box

WL cont, WL int, QW cont, QW int, PR cont, PR int, EV cont, EV int, wind vel., tide cont, tide int, sed. con, sed. ps, peak flow, low flow, state water use

INSTRUMENTS (C805)
(Place a "Y" in the appropriate box):

digital rec-order, graphic rec-order, tele-metry land line, tele-metry radio, tele-metry satellite, AHDAS, crest-stage gage, tide gage, deflec-tion meter, bubble gage, stilling well, CR type recorder, weigh-ing rain gage, tipping bucket rain gage, acoustic velocity meter, electro-magnetic flowmeter, pressure transducer

DATE INVENTORIED (C711) month - day - year

RECORD READY FOR WEB (C32): Y ready to display, C condi-tional, P proprie-tary, L local use only

REMARKS (C806)

FOOTNOTES

1 SITE TYPE (C802)

GL	Glacier	OC	Ocean	GW	Well	SB	Subsurface
WE	Wetland	OC-CO	Coastal	GW-CR	Collector or Ranney type well	SB-CV	Cave
AT	Atmosphere	LK	Lake, Reservoir, Impoundment	GW-EX	Extensometer well	SB-GWD	Groundwater drain
ES	Estuary			GW-HZ	Hyporheic -zone well	SB-TSM	Tunnel, shaft, or mine
LA	Land	SP	Spring	GW-IW	Interconnected wells	SB-UZ	Unsaturated zone
LA-EX	Excavation	ST	Stream	GW-TH	Test hole not completed as a well		
LA-OU	Outcrop	ST-CA	Canal	GW-MW	Multiple wells		
LA-SNK	Sinkhole	ST-DCH	Ditch				
LA-SH	Soil hole	ST-TS	Tidal stream				
LA-SR	Shore	FA-WIW	Waste-Injection well				

2 WS water supply, DO domestic, CO commer-cial, IN industrial, IR irrigation, MI mining, LV livestock, PH power hydro-electric, ST waste water treatment, RM remedia-tion, TE thermo-electric power, AQ aqua-culture

C22 Other (see manual for codes)
C36 Other (see manual for codes)
C39 is mandatory for all sites having data in SWUDS.

Figure 2. Groundwater Site Schedule, Form 9-1904-A.

GENERAL SITE DATA

DATA RELIABILITY (C3)

C	L	M	U
field checked	poor location	minimal data	un-checked

DATE OF FIRST CONSTRUCTION (C21) ☐☐ – ☐☐ – ☐☐☐☐ (month – day – year)

USE OF SITE (C23)

A	C	D	E	G	H	M	O	P	R	S	T	U	V	W	X	Z
anode	standby emer. supply	drain	geo-thermal	seismic	heat reservoir	mine	obser-vation	oil or gas	recharge	repres-surize	test	unused	with-drawal/ return	with-drawal	waste	des-troyed

SECOND-ARY USE OF SITE (C301) (See use of site) ☐

TERTIARY USE OF SITE (C302) (See use of site) ☐

USE OF WATER (C24)

A	B	C	D	E	F	H	I	J	K	M	N	P	Q	R	S	T	U	Y	Z
air cond.	bottling	comm-ercial	de-water	power	fire	domes-tic	irri-gation	indus-trial (cooling)	mining	medi-cinal	indus-trial	public supply	aqua-culture	recrea-tions	stock	insti-tutional	unused	desalin-ation	other

SECOND-ARY USE OF WATER (C25) (see use of water) ☐

TERTIARY USE OF WATER (C26) (see use of water) ☐

AQU FER TYPE (C713)

U	N	C	M	X
unconfined single	unconfined multiple	confined single	confined multiple	mixed

PR MARY AQUIFER (C714) ☐☐☐☐☐☐☐☐

NATIONAL AQU FER (C715) ☐☐☐☐☐☐☐☐☐☐

HOLE DEPTH (C27) ☐☐☐☐☐ . ☐☐

WELL DEPTH (C28) ☐☐☐☐☐ . ☐☐

SOURCE OF DEPTH DATA (C29)

A	D	G	L	M	O	R	S	Z
other gov't	driller	geol-ogist	logs	memory	owner	other reported	reporting agency	other

WATER-LEVEL DATA

DATE WATER-LEVEL MEASURED (C235) ☐☐ – ☐☐ – ☐☐☐☐ (month – day – year)

T ME (C709) ☐☐☐☐

WATER-LEVEL TYPE CODE (C243)

L	M	S
land surface	meas. pt.	vertical datum

WATER LEVEL (C237/241/242) ☐☐☐☐☐ . ☐☐

MP SEQUENCE NO. (C248) (Mandatory if WL type=M) ☐☐☐

WATER-LEVEL DATUM (C245) (Mandatory if WL type=S)

NGVD29	NAVD88	☐☐☐☐☐☐☐☐☐☐
National Geodetic Vertical Datum 0f 1929	North American Vertical Datum 0f 1988	Other (See manual for codes)

SITE STATUS FOR WATER LEVEL (C238)

A	B	C	D	E	F	G	H	I	J	M	N	O	P	R	S	T	V	W	X	Z
atmos. pressure	tide stage	ice	dry	recently flowing	flowing	nearby flowing	nearby recently flowing	injector site	injector site monitor	plugged	measure-ment discontinued	obstruc-tion	pumping	recently pumped	nearby pumping	nearby recently pumped	foreign sub-stance	well des-troyed	affected by surface water	other

METHOD OF WATER-LEVEL MEASUREMENT(C239)

A	B	C	D	E	F	G	H	L	M	N	O	P	R	S	T	V	Z
airline	analog	calibrated airline	differ-ential GPS	esti-mated	trans-ducer	pressure gage	calibrated press. gage	geophysi-cal logs	mano-meter	non-rec. gage	observed	acoustic pulse	reported	steel tape	electric tape	calibrated elec. tape	other

WATER-LEVEL ACCURACY (C276)

0	1	2	9
foot	tenth	hun-dredth	not to nearest foot

SOURCE OF WATER-LEVEL DATA (C244)

A	D	G	L	M	O	R	S	Z
other gov't	driller's log	geol-ogist	geophysi-cal logs	memory	owner	other reported	reporting agency	other

PERSON MAKING MEASUREMENT (C246) (WATER LEVEL PARTY) ☐☐☐☐☐☐

MEASURING AGENCY (C247) (SOURCE) ☐☐☐☐☐

EQU P D (C249) (20 char) ____________________

REMARKS (C267) (256 char) ____________________

RECORD READY FOR WEB (C858)

Y	C	P	L
ready to display	condi-tional	proprie-tary	local use only

CONSTRUCTION DATA

RECORD TYPE (C754) CONS

RECORD SEQUENCE NO. (C723) ☐☐☐

DATE OF COMPLETED CONSTRUCTION (C60) ☐☐ – ☐☐ – ☐☐☐☐ (month – day – year)

NAME OF CONTRACTOR (C63) ☐☐☐☐☐☐☐☐☐☐☐☐

SOURCE OF DATA (C64)

A	D	G	L	M	O	R	S	Z
other gov't	driller	geol-ogist	logs	memory	owner	other reported	reporting agency	other

METHOD OF CONSTRUCTION (C65)

A	B	C	D	H	J	P	R	S	T	V	W	Z
air-rotary	bored or augered	cable tool	dug	hydraulic rotary	jetted	air per-cussion	reverse rotary	sonic	trenching	driven	drive wash	other

TYPE OF FINISH (C66)

C	F	G	H	O	P	S	T	W	X	Z
porous concrete	gravel w/perf.	gravel screen	horiz. gallery	open end	perf or slotted	screen	sand point	walled	open hole	other

TYPE OF SEAL (C67)

B	C	G	N	Z
bentonite	clay	cement grout	none	other

BOTTOM OF SEAL (C68) ☐☐☐☐

METHOD OF DEVELOPMENT (C69)

A	B	C	J	N	P	S	Z
air-lift pump	bailed	compres-sed air	jetted	none	pumped	surged	other

HOURS OF DEVELOPMENT (C70) ☐☐☐

SPECIAL TREATMENT (C71)

C	D	E	F	H	M	Z
chem-icals	dry ice	explo-sives	defloc-culent	hydro-frac-turing	mech-anical	other

CONSTRUCTION HOLE DATA (3 sets shown)

RECORD TYPE (C756) H O L E
RECORD SEQUENCE NO. (C724) SEQUENCE NO. OF PARENT RECORD (C59)

DEPTH TO TOP OF INTERVAL (C73) . DEPTH TO BOTTOM OF NTERVAL (C74) . DIAMETER OF INTERVAL (C75) .

RECORD SEQUENCE NO. (C724)

DEPTH TO TOP OF INTERVAL (C73) . DEPTH TO BOTTOM OF NTERVAL (C74) . DIAMETER OF INTERVAL (C75) .

RECORD SEQUENCE NO. (C724)

DEPTH TO TOP OF INTERVAL (C73) . DEPTH TO BOTTOM OF NTERVAL (C74) . DIAMETER OF INTERVAL (C75) .

CONSTRUCTION CASING DATA (4 sets shown)

RECORD TYPE (C758) C S N G
RECORD SEQUENCE NO. (C725) SEQUENCE NO. OF PARENT RECORD (C59)

DEPTH TO TOP OF CAS NG (C77) . DEPTH TO BOTTOM OF CASING (C78) . DIAMETER OF CASING (C79) .

4 CASING MATERIAL (C80) CASING THICKNESS (C81) .

RECORD SEQUENCE NO. (C725) SEQUENCE NO. OF PARENT RECORD (C59)

DEPTH TO TOP OF CASING (C77) . DEPTH TO BOTTOM OF CASING (C78) . DIAMETER OF CASING (C79) .

4 CASING MATERIAL (C80) CASING THICKNESS (C81) .

RECORD SEQUENCE NO. (C725) SEQUENCE NO. OF PARENT RECORD (C59)

DEPTH TO TOP OF CAS NG (C77) . DEPTH TO BOTTOM OF CASING (C78) . DIAMETER OF CASING (C79) .

4 CAS NG MATERIAL (C80) CAS NG THICKNESS (C81) .

RECORD SEQUENCE NO. (C725) SEQUENCE NO. OF PARENT RECORD (C59)

DEPTH TO TOP OF CAS NG (C77) . DEPTH TO BOTTOM OF CASING (C78) . DIAMETER OF CASING (C79) .

4 CAS NG MATERIAL (C80) CAS NG THICKNESS (C81) .

FOOTNOTE:

4 CAS NG MATERIAL CODES

A	B	C	D	E	F	G	H	I	J	K	L	M	N	P	Q	R	S	T	U	V	W	X	Y	Z	4	6
abs	brick	concrete	copper	PTFE	Fiber-glass	galv. iron	Fiber-glass plastic	wrought iron	Fiber-glass epoxy	PVC thread-ed	glass	other metal	PVC glued	PVC or plastic	FEP	rock or stone	steel	tile	coated steel	stain-less steel	wood	steel carbon	steel galva-nized	other mat.	stain-less 304	stain-less 316

CONSTRUCTION OPENINGS DATA (3 sets shown)

RECORD TYPE (C760) OPEN RECORD SEQUENCE NO. (C726) SEQUENCE NO. OF PARENT RECORD (C59)

DEPTH TO TOP OF INTERVAL (C83) . DEPTH TO BOTTOM OF INTERVAL (C84) . DIAMETER OF INTERVAL (C87) .

5 MATERIAL TYPE (C86) 6 TYPE OF OPENING (C85) LENGTH OF OPENING (C89) . WIDTH OF OPENING (C88) .

RECORD SEQUENCE NO. (C726)

DEPTH TO TOP OF INTERVAL (C83) . DEPTH TO BOTTOM OF INTERVAL (C84) . DIAMETER OF INTERVAL (C87) .

5 MATERIAL TYPE (C86) 6 TYPE OF OPENING (C85) LENGTH OF OPENING (C89) . WIDTH OF OPENING (C88) .

RECORD SEQUENCE NO. (C726)

DEPTH TO TOP OF INTERVAL (C83) . DEPTH TO BOTTOM OF INTERVAL (C84) . DIAMETER OF INTERVAL (C87) .

5 MATERIAL TYPE (C86) 6 TYPE OF OPENING (C85) LENGTH OF OPENING (C89) . WIDTH OF OPENING (C88) .

FOOTNOTES:

5 TYPE OF MATERIAL CODES FOR OPEN SECTIONS

A	B	C	D	E	F	G	H	I	J	K	L	M	N	P	Q	R	S	T	V	W	X	Y	Z	4	6
ABS	brass or bronze	concrete	ceramic	PTFE	fiber-glass	galv. iron	fiber-glass plastic	wrought iron	fiber-glass epoxy	PVC threaded	glass	other metal	PVC glued	PVC	FEP	stain-less steel	steel	tile	brick	mem-brane	steel carbon	steel galva-nized	other	stain-less 304	stain-less 316

6 TYPE OF OPENINGS CODES

F	L	M	P	R	S	T	W	X	Z
fractured rock	louvered or shutter-type	mesh screen	perforated, porous or slotted	wire-wound screen	screen (unk.)	sand point screen	walled or shored	open hole	other

CONSTRUCTION MEASURING POINT DATA

RECORD TYPE (C766) MPNT RECORD SEQUENCE NO. (C728) BEGINNING DATE (C321) – – month day year ENDING DATE (C322) – –

M.P. HEIGHT (C323) . ALTITUDE OF MEASURING POINT (C325) ALTITUDE METHOD (C326) ALTITUDE ACCURACY (C327)

ALTITUDE DATUM (C328) M.P. REMARKS (C324)

RECORD READY FOR WEB (C857)

Y	C	P	L
ready to display	condi-tional	proprie-tary	local use only

CONSTRUCTION LIFT DATA

RECORD TYPE (C752) L I F T

RECORD SEQUENCE NO. (C254)

TYPE OF LIFT (C43): A air, B bucket, C centri-fugal, J jet, P piston, R rotary, S submer-sible, T turbine, U un-known, X no lift, Z other

DATE RECORDED (C38) month – day – year

PUMP INTAKE DEPTH (C44)

TYPE OF POWER (C45): D diesel, E electric, G gaso-line, H hand, L LP gas, N natural gas, S solar, W windmill, Z other

HORSE-POWER RATING (C46) .

MANUFACTURER (C48)

SERIAL NO. (C49)

POWER COMPANY (C50)

POWER COMPANY ACCOUNT NUMBER (C51)

POWER METER NUMBER (C52)

PUMP RATING (C53) (million gallons/units of fuel) .

ADDITIONAL LIFT (C255)

PERSON OR COMPANY MAINTAINING PUMP (C54)

RATED PUMP CAPACITY (gpm) (C268)

STANDBY POWER (C56) (see TYPE OF POWER)

HORSEPOWER OF STANDBY POWER SOURCE (C57) .

MISCELLANEOUS OWNER DATA

RECORD TYPE (C768) O W N R

RECORD SEQUENCE NO. (C718)

DATE OF OWNERSH P (C159) – –

WU OWNER TYPE (C350): CP Corporation, GV Govern-ment, IN Individual, MI Military, OT Other, TG Tribal, WS Water Supplier

END DATE OF OWNERSHIP (C374) – –

OWNER'S NAME (C161)

EXAMPLES: JONES, RALPH A.
JONES CONSTRUCTION COMPANY

OWNER'S PHONE NUMBER (C351)

ACCESS TO OWNER'S NAME (C352): 0 Public Access, 1 Coop-erator, 2 USGS Only, 3 District Only, 4 Proprietary

OWNER'S ADDRESS (LINE 1) (C353)

OWNER'S ADDRESS (LINE 2) (C354)

OWNER'S CITY NAME (C355)

STATE (C356)

OWNER'S Z P CODE (C357) –

OWNER'S COUNTRY NAME (C358)

ACCESS TO OWNER'S PHONE/ADDRESS (C359): 0 Public Access, 1 Coop-erator, 2 USGS Only, 3 District Only, 4 Proprietary

MISCELLANEOUS VISIT DATA

RECORD TYPE (C774) V I S T

RECORD SEQUENCE NO. (C737)

DATE OF VISIT (C187) month – day – year

NAME OF PERSON (C188)

MISCELLANEOUS OTHER ID DATA (2 sets shown)

RECORD TYPE (C770) O T I D

RECORD SEQUENCE NO. (C736) | OTHER ID (C190)

ASSIGNER (C191)

RECORD SEQUENCE NO. (C736) | OTHER ID (C190)

ASSIGNER (C191)

MISCELLANEOUS OTHER DATA

RECORD TYPE (C772) O T D T | RECORD SEQUENCE NO. (C312)

OTHER DATA TYPE (C181)

OTHER DATA LOCATION (C182): C Cooperator's Office, D District Office, R Reporting Agency, Z other

DATA FORMAT (C261): F files, M machine readable, P published, Z other

MISCELLANEOUS LOGS DATA (3 sets shown)

RECORD TYPE (C778) L O G S | RECORD SEQUENCE NO. (C739) | TYPE OF LOG (C199)

BEG NN NG DEPTH (C200) | ENDING DEPTH (C201) | SOURCE OF DATA (C202): A other gov't, D driller, G geol-ogist, L logs, M memory, O owner, R other reported, S reporting agency, Z other

DATA FORMAT (C225): F files, M machine readable, P published, Z other | OTHER DATA LOCATION (C226)

RECORD TYPE (C778) L O G S | RECORD SEQUENCE NO. (C739) | TYPE OF LOG (C199)

BEG NNING DEPTH (C200) | ENDING DEPTH (C201) | SOURCE OF DATA (C202): A other gov't, D driller, G geol-ogist, L logs, M memory, O owner, R other reported, S reporting agency, Z other

DATA FORMAT (C225): F files, M machine readable, P published, Z other | OTHER DATA LOCATION (C226)

RECORD TYPE (C778) L O G S | RECORD SEQUENCE NO. (C739) | TYPE OF LOG (C199)

BEGINNING DEPTH (C200) | ENDING DEPTH (C201) | SOURCE OF DATA (C202): A other gov't, D driller, G geol-ogist, L logs, M memory, O owner, R other reported, S reporting agency, Z other

DATA FORMAT (C225): F files, M machine readable, P published, Z other | OTHER DATA LOCATION (C226)

ACOUSTIC LOG:
AS Sonic
AV Acoustic velocity
AW Acoustic waveform
AT Acous ic televiewer

CALIPER LOG:
CP Caliper
CS Caliper, single arm
CT Caliper, three arm
CM Caliper, multi arm
CA Caliper, acous ic

DRILLING LOG:
DT Drilling ime
DR Drillers
DG Geologists
DC Core

ELECTRIC LOG:
EE Electric
ER Single-point resistance
EP Spontaneous potential
EL Long-normal resistivity
ES Short-normal resis ivity
EF Focused resistivity
ET Lateral resistivity
EN Microresistivity
EC Microresistivity, forused
EO Microresis ivity, lateral
ED Dipmeter

ELECTROMAGNETIC LOG:
MM Magnetic log
MS Magnetic susceptibiity log
MI Electromagne ic induc ion log
MD Electromagne ic dual induction log
MR Radar reflection image log
MV Radar direct-wave velocity log
MA Radar direct-wave amplitude log

FLUID LOG:
FC Fluid conductivity
FR Fluid resistivity
FT Fluid temperature
FF Fluid differen ial temperature
FV Fluid velocity
FS Spinner flowmeter
FH Heat-pulse flowmeter
FE Electromagnetic flowmeter
FD Doppler flowmeter
FA Radioactive tracer
FY Dye tracer
FB Brine tracer

NUCLEAR LOG:
NG Gamma
NS Spectral gamma
NA Gamma-gamma
NN Neutron
NT Neutron activitation
NM Neuclear magnetic resonance

OPTICAL LOG:
OV Video
OF Fisheye video
OS Sidewall video
OT Optical televiewer

COMBINATION LOG:
ZF Gamma, fluid resistivity, temperature
ZI Gamma, electromagne ic induction
ZR Long/short normal resistivity
ZT Fluid resis ivity, temperature
ZM Electromagnetic flowmeter, fluid resistivity, temperature
ZN Long/short normal resistivity, spontaneous poten ial
ZP Single-point resistance, spontaneous potential
ZE Gamma, long/short normal resistivity, spontaneous potential, single-point resistance, fluid resi ivity, temperature

WELL CONSTRUCTION LOG:
WC Casing collar
WD Borehold deviation

OTHER LOG:
OR Other

MISCELLANEOUS NETWORK DATA (3 types shown)

RECORD TYPE (C780) N E T W — RECORD SEQUENCE NO. (C730) — TYPE OF NETWORK (C706) Q W (water quality) — BEGINNING YEAR (C115) — ENDING YEAR (C116)

TYPE OF ANALYSIS (C120)

A	B	C	D	E	F	G	H	I	J	K	L	M	N	P	Z
physical proper-ties	common ions	trace elements	pesti-cides	nutri-ents	sanitary analysis	codes D&B	codes B&E	codes B&C	codes B&F	codes D&E	codes C,D&E	all or most	codes B&C& radio-active	codes B,C&A	other

SOURCE AGENCY (C117) — [7] FREQUENCY OF COLLECTION (C118) — ANALYZING AGENCY (C307) — [8] PRIMARY NETWORK SITE (C257) — [8] SECONDARY NETWORK SITE (C708)

RECORD TYPE (C780) N E T W — RECORD SEQUENCE NO. (C730) — TYPE OF NETWORK (C706) W L (water level) — BEGINNING YEAR (C115) — ENDING YEAR (C116)

SOURCE AGENCY (C117) — [7] FREQUENCY OF COLLECTION (C118) — [8] PRIMARY NETWORK SITE (C257) — [8] SECONDARY NETWORK SITE (C708)

RECORD TYPE (C780) N E T W — RECORD SEQUENCE NO. (C730) — TYPE OF NETWORK (C706) W D (pumpage or with-drawals) — BEGINNING YEAR (C115) — ENDING YEAR (C116)

SOURCE AGENCY (C117) — [7] FREQUENCY OF COLLECTION (C118) — METHOD OF COLLECTION (C133) — [8] PRIMARY NETWORK SITE (C257) — [8] SECONDARY NETWORK SITE (C708)

C	E	M	U	Z
calcu-lated	esti-mated	meter-ed	un-known	other

FOOTNOTES:

[7] FREQUENCY OF COLLECTION CODES

A	B	C	D	F	I	M	O	Q	S	W	Z	2	3	4	5	X
annually	bi monthly	continu-ously	daily	semi-monthly	inter mittent	monthly	one-time only	quarter-ly	semi-annually	weekly	other	bi-annually	every 3 years	every 4 years	every 5 years	every 10 years

[8] NETWORK SITE CODES

1	2	3	4
national,	district,	project,	co-operator,

MISCELLANEOUS REMARKS DATA (4 types shown)

RECORD TYPE (C788) R M K S — RECORD SEQUENCE NO. (C311) — DATE OF REMARK (C184) month – day – year

REMARKS (C185)

Subsequent entries may be used to continue the remark. Miscellaneous remarks field is limited to 256 characters.

RECORD TYPE (C788) R M K S — RECORD SEQUENCE NO. (C311) — DATE OF REMARK (C184) month – day – year

REMARKS (C185)

Subsequent entries may be used to continue the remark. Miscellaneous remarks field is limited to 256 characters.

DISCHARGE DATA

RECORD SEQUENCE NO. (C147)

DATE DISCHARGE MEASURED (C148) — month – day – year

TYPE OF DISCHARGE (C703) P pumped F flow

DISCHARGE (gpm) (C150)

ACCURACY OF DISCHARGE MEASUREMENT (C310): E excellent (LT 2%), G good (2%-5%), F fair (5%-8%), P poor (GT 8%)

SOURCE OF DATA (C151): A other gov't, D driller, G geologist, L logs, M memory, O owner, R other reported, S reporting agency, Z other

METHOD OF DISCHARGE MEASUREMENT (C152): A acoustic meter, B bailer, C current meter, D Doppler meter, E estimated, F flume, M totaling meter, O orifice, P pitot-tube, R reported, T trajectory, U venturi meter, V volumetric meas, W weir, X unknown, Z other

PRODUCTION WATER LEVEL (C153)

STATIC WATER LEVEL (C154)

SOURCE OF DATA (C155): A other gov't, D driller, G geologist, L logs, M memory, O owner, R other reported, S reporting agency, Z other

METHOD OF WATER-LEVEL MEASUREMENT (C156): A airline, B recorder, C calibrated airline, D differential GP, E estimated, F transducer, G pressure gage, H calibrated press. gage, L geophysical logs, M manometer, N non-rec. gage, O observed, P acoustic pulse, R reported, S steel tape, T electric tape, V calibrated elec. tape, Z other

PUMPING PERIOD (C157)

SPECIFIC CAPACITY (C272)

DRAWDOWN (C309)

GEOHYDROLOGIC DATA

RECORD TYPE (C748) GEOH

RECORD SEQUENCE NO. (C721)

DEPTH TO TOP OF UNIT (C91)

DEPTH TO BOTTOM OF UNIT (C92)

UNIT IDENTIFIER (C93)

LITHOLOGY (C96)

CONTRIBUTING UNIT (C304): P principal aquifer, Q aggregate of lithologic units, S secondary aquifer, N no contribution, U unknown

LITHOLOGIC MOD FIER (C97)

GEOHYDROLOGIC AQUIFER DATA

RECORD TYPE (C750) AQFR

RECORD SEQUENCE NO. (C742)

SEQUENCE NO. OF PARENT RECORD (C256)

DATE (C95) month – day – year

STATIC WATER LEVEL (C126)

CONTRIBUTION (C132)

SITE LOCATION SKETCH AND DIRECTIONS

Township ___________ Range ___________

Section #___________

GWPD 6—Recognizing and removing debris from a well

VERSION: 2010.1

PURPOSE: To recognize when a well contains debris and how to remove the debris from the well.

Materials and Instruments

1. Steel tape graduated in feet, tenths and hundredths of feet, or an electric tape
2. Blue carpenter's chalk
3. Clean rag
4. Mirror
5. Flashlight
6. Pencil or pen, blue or black ink. Strikethrough, date, and initial errors; no erasures
7. Field notebook
8. Water-level measurement field form or handheld computer for data entry
9. A grappling device with wire line or heavy duty treble fishing hook and rope
10. Safety equipment: gloves, safety glasses, first-aid kit

Data Accuracy and Limitations

1. Debris that is present in a well can affect the plumbness of the tape and cause errors in water-level measurements.
2. The quality of water-level data from a well is directly related to well maintenance.
3. Success rate for this procedure increases with increasing well diameter and decreasing well depth.

Assumptions

1. Individual has been trained to make water-level measurements with a graduated steel tape (GWPD 1) or an electric tape (GWPD 4).
2. State or local ordinances do not prevent retrieval of an item in a well.

Instructions

1. Make a water-level measurement as described in GWPD 1 or GWPD 4. Lack of agreement between the original water-level measurement and subsequent water-level check measurements could indicate that the well contains debris. If the measuring tape goes slack as it is being slowly lowered into the well, the weight or probe probably has encountered debris in the well.
2. To check for debris on a sunny day, use a mirror to look into the well. Hold the mirror in the hand and rotate it back and forth until the proper angle is obtained to allow the sun to reflect off the mirror and down the well onto the water surface.
3. If the well is located in a dark enclosed area away from the sun, or the weather is overcast, use a flashlight to look down the well for debris.
4. To remove light- to medium-weight wood debris from a well, use a simple inexpensive device such as a heavy duty treble fishing hook attached to a rope. Lower the hook down the well while using the mirror to see when the hook is below the debris. To remove the debris from the well, move the rope upward with a quick jerking motion until the wood debris becomes snagged on the treble hook. Slowly remove the rope and debris from the well. If the object is below the water surface where it cannot be seen, feel for the debris while trying to snag it.

5. To remove heavy wood or debris that cannot be snagged, use a grappling device similar to a pair of ice tongs. The device shown in figure 1 has been designed and used to remove debris from wells effectively and easily. This type of device can be used to remove blocks of wood, stones, cans, bottles, pipes, and poles from wells and can be constructed by a machine shop from the photographs shown in figure 1. To remove debris from a well, cock the device in the open position (fig. 1*B*) and lower into the well on a suspension cable that is fastened to a shackle. When the tripping rod strikes the debris in the well, the rod pushes upward on the locking bar, releasing it, and the spring opposite the locking bar (fig. 1*B*) pulls the arms together. Figure 1*C* shows the grappling device in the closed position gripping a heavy object (15 pounds). The weight of the debris being lifted from the well holds the arms together. The heavier the object, the tighter the arms grip. In case the tripping rod will not close the arms, the arms can be closed from the surface by attaching a line at the pivot point of the locking bar. Lower the grappling device into the well and pull on the line connected to the locking bar when the arms are in the desired position. The arms will close around the debris without the aid of the tripping rod.

Data Recording

Data are recorded in a field notebook and on a water-level measurement field form (fig. 2).

A. Device in open position

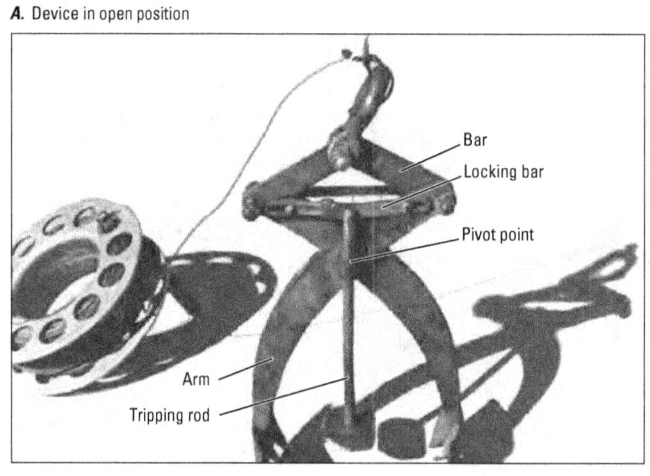

B. Detailed view of locking bar and releasing rod

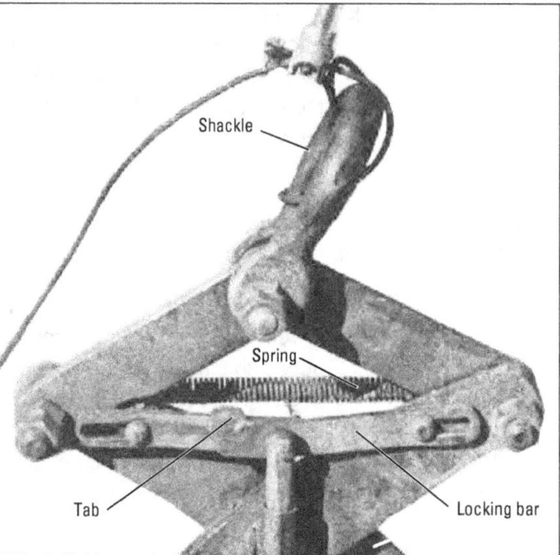

C. Device in closed position

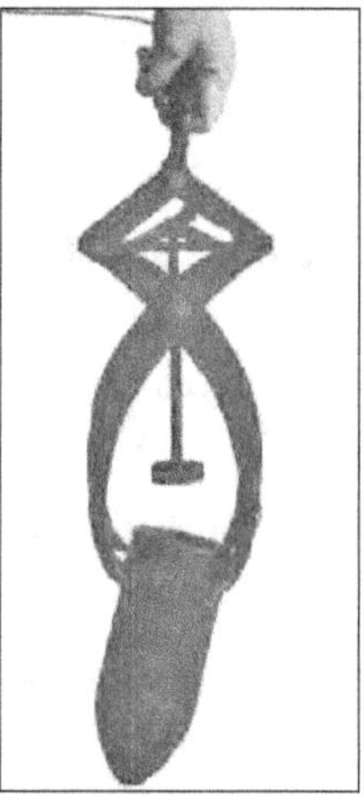

Figure 1. Grappling device for removing debris from wells (Bader, 1966).

WATER-LEVEL MEASUREMENT FIELD FORM
Steel Tape Measurement

SITE INFORMATION

SITE ID (C1) ____

Equipment ID ____ Date of Field Visit ____

Station name (C12) ____

WATER-LEVEL DATA

	1	2	3	4	5
Time					
Hold					
Cut					
Tape correction					
WL below MP					
MP correction					
WL below LSD					

Measured by ____ COMMENTS* ____

*Comments should include quality concerns and changes in: M.P., ownership, access, locks, dogs, measuring problems, et al.

MEASURING POINT DATA (for MP Changes)

M.P. REMARKS (C324) | BEGINNING DATE (C321) month-day-year | ENDING DATE (C322) | M.P. HEIGHT (C323) NOTE: (-) for MP below land surface

Final Measurement for GWSI

WATER LEVEL TYPE CODE (C243): L below land surface; M below meas. pt.; S sea level

DATE WATER LEVEL MEASURED (C235) month-day-year | TIME (C709) | STATUS (C238) | METHOD (C239) | TYPE (C243) | WATER LEVEL (C237)

METHOD OF WATER-LEVEL MEASUREMENT(C239)

A	B	C	E	G	H	L	M	N	R	S (GWPD1)	T	V (GWPD4)	Z
airline,	analog,	calibrated airline,	estimated,	pressure gage,	calibrated press. gage,	geophysical logs,	manometer,	non-rec. gage,	reported,	steel tape,	electric tape,	calibrated elec. tape	other

SITE STATUS FOR WATER LEVEL (C238)

D	E	F	G	H	I	J	M	N	O	P	R	S	T	V	W	X	Z	BLANK
dry,	recently flowing,	flowing,	nearby flowing	nearby recently flowing,	injector site,	injector site monitor,	plugged,	measurement discon.,	obstruction,	pumping,	recently pumped,	nearby pumping,	nearby recently pumped,	foreign substance,	well destroyed,	surface water effects,	other	static

Figure 2. Water-level measurement field form for steel tape measurements. This form, or an equivalent custom-designed form, should be used to record field measurements.

References

Bader, J.S., 1966, Device for removing debris from wells, *in* Mesnier, G.N., and Chase, E.B., comps., Selected techniques in water resources investigations, 1965: U.S. Geological Survey Water-Supply Paper 1822, p. 43–46.

Cunningham, W.L., and Schalk, C.W., comps., 2011a, Groundwater technical procedures of the U.S. Geological Survey, GWPD 1—Measuring water levels by use of a graduated steel tape: U.S. Geological Survey Techniques and Methods 1–A1, 4 p.

Cunningham, W.L., and Schalk, C.W., comps., 2011b, Groundwater technical procedures of the U.S. Geological Survey, GWPD 4—Measuring water levels by use of an electric tape: U.S. Geological Survey Techniques and Methods 1–A1, 6 p.

GWPD 7—Estimating discharge from a naturally flowing well

VERSION: 2010.1

PURPOSE: To estimate the discharge from a naturally flowing well from a vertical pipe.

Materials and Instruments

1. Small hand level
2. L-shaped measuring device (carpenter's square), graduated by inches
3. Clamp
4. Support rod for the measuring device
5. Field notebook
6. Pencil or pen, blue or black ink. Strikethrough, date, and initial errors; no erasures
7. Ground-Water Site-Inventory (GWSI) System Groundwater Site Schedule, Form 9-1904-A

Data Accuracy and Limitations

1. Under ordinary field conditions, with reasonable care, measurements may be made in which the error seldom exceeds 10 percent.
2. Not accurate for small flows of 30 gallons per minute or less, or when the crest of the flow is less than 1.5 inches. For small flows, connect a pipe tee to the top of the well casing and measure the well discharge with a bucket and stopwatch.
3. The most accurate estimated discharge will be obtained when the pipe is truly vertical.

Advantages

1. Fast and simple means of approximating the flow from vertical pipes.
2. No special training needed to use this method.

Disadvantages

1. Method provides only an approximate discharge from wells with vertical pipes.
2. Well flow must be constant so that the height of water above the pipe does not vary appreciably.

Assumptions

1. The discharge pipe does not have a circular orifice weir.
2. The discharge pipe does not have an in-line flowmeter.
3. The pipe is vertical.

Instructions

1. Measure the height of the crest of the water flow, in inches, above the top of the vertical pipe. This measurement can be made using a small hand level, an L-shaped measuring device, a clamp, and a support rod. Figure 1 shows how to set up the equipment to measure the height of the crest of flow from a vertical pipe.

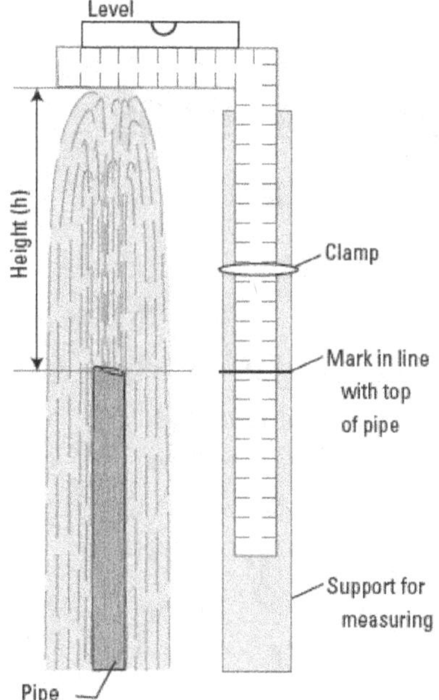

Figure 1. Measuring the height of the crest of flow from a vertical pipe. (Driscoll, 1966, p. 97)

2. Measure the inside diameter of the discharge pipe, in inches.

3. Estimate well discharge from the discharge curves shown in figure 2 for vertical standard pipes. Find the number that corresponds to the height of the crest of the water flow on the y-axis. Move horizontally to the right along that line to the curve that represents the inside diameter of the well. Read the discharge, in gallons per minute, from the x-axis corresponding to that point. If the inside diameter of the well for which discharge is being estimated is not one of the given curves in figure 2, estimate the well discharge by interpolating between the curves. Read the discharge, in gallons per minute, and record the results in the field notebook and in the discharge data section of the GWSI Groundwater Site Schedule (fig. 3, Form 9-1904-A).

Data Recording

Data are recorded in a field notebook. Discharge data also should be recorded in the discharge data section of the GWSI Groundwater Site Schedule (fig. 3, Form 9-1904-A). This is best described as a trajectory method and should be coded as "T" in field C152 on Form 9-1904-A.

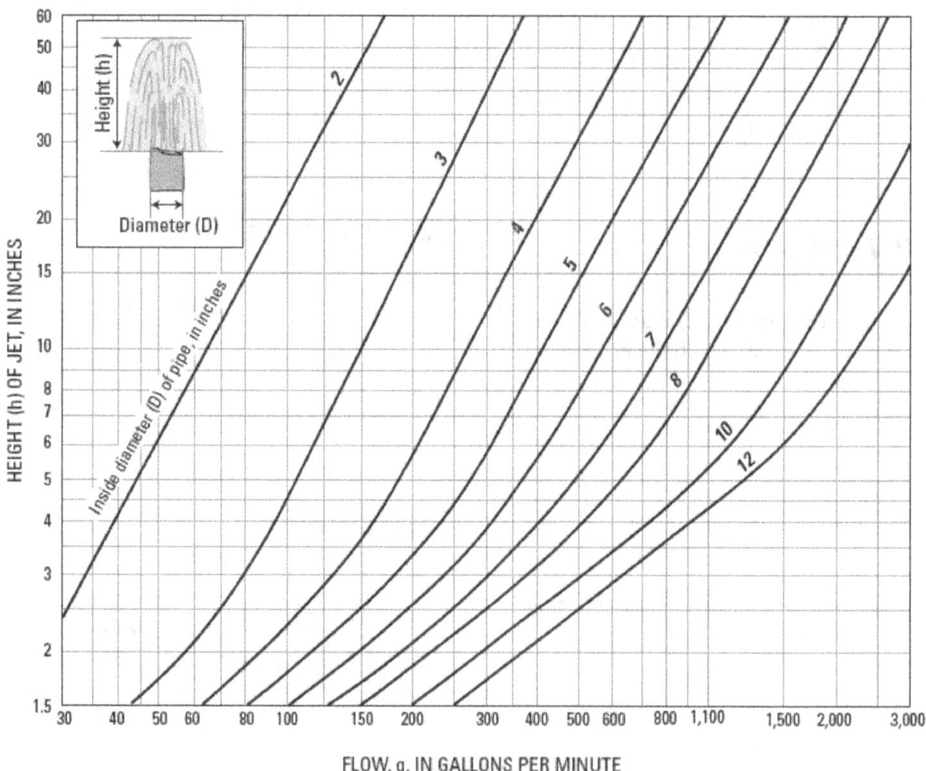

Figure 2. Discharge curves for measurement of flow from vertical standard pipes. The curves are based on data from experiments of Lawrence and Braunworth (1906). (From Bureau of Reclamation. 1967, p. 199)

FORM NO. 9-1904-A
Revised Sept 2009, NWIS 4.9

File Code ____________
Date ____________

Coded by ____________
Checked by ____________
Entered by ____________

U.S DEPT. OF THE INTERIOR
GEOLOGICAL SURVEY

GROUNDWATER SITE SCHEDULE
General Site Data

AGENCY CODE (C4) USGS | SITE ID (C1) | PROJECT (C5)

STATION NAME (C12/900)

SITE TYPE[1] (C802) Primary - Secondary | DISTRICT (C6) | COUNTRY (C41) | STATE (C7)

COUNTY or TOWN (C8) ____________ County code

LATITUDE (C9) | LONGITUDE (C10) | LAT/LONG ACCURACY (C11): H Hndrth sec. 1 tenth sec. 5 half sec. S sec. R 3 sec. F 5 sec. T 10 sec. M min. U Un-known

LAT/LONG METHOD (C35): C land net, D DGPS, G GPS, L LORAN, M map, N Inter-polated digital map, R reported, S survey, U un-known

LAT/LONG DATUM (C36): NAD27 North American Datum of 1927, NAD83 North American Datum of 1983

ALTITUDE (C16)

ALTITUDE ACCURACY (C18) | ALTITUDE METHOD (C17): A altimeter, D DGPS, G GPS, I InSAR, J LIDAR, L Level, M map, N DEM, R re-ported, U un-known

ALTITUDE DATUM (C22): NGVD29 National Geodetic Vertical Datum of 1929, NAVD88 North American Vertical Datum of 1988

LAND NET (C13): ¼ ¼ ¼ S section T township range merid

TOPOGRAPHIC SETTING (C19): A alluvial fan, B playa, C stream channel, D depres-sion, E dunes, F flat, G flood-plain, H hill-top, K sink-hole, L lake or swamp, M mangrove swamp, O off-shore, P pedi-ment, S hill-side, T ter-race, U undu-lating, V valley flat, W upland draw

HYDROLOGIC UNIT CODE (C20) | DRAINAGE BASIN CODE (C801) | STANDARD TIME ZONE (C813) | DAYLIGHT SAVINGS TIME FLAG (C814) Y OR **N**

MAP NAME (C14) | MAP SCALE (C15)

AGENCY USE (C803): A active no/na, D discon-tinued, I inactive site, L active written, M active oral, O inventory site, R remediated

2 NATIONAL WATER-USE (C39)

DATA TYPE (C804) Place an 'A' (active), an 'I' (inactive), or an 'O' (inventory) in the appropriate box: WL cont, WL int, QW cont, QW int, PR cont, PR int, EV cont, EV int, wind vel., tide cont, tide int, sed. con, sed. ps, peak flow, low flow, state water use

INSTRUMENTS (C805) (Place a "Y" in the appropriate box): digital rec-order, graphic rec-order, tele-metry land line, tele-metry radio, tele-metry satellite, AHDAS, crest-stage gage, tide gage, deflec-tion meter, bubble gage, stilling well, CR type recorder, weigh-ing rain gage, tipping bucket rain gage, acoustic velocity meter, electro-magnetic flowmeter, pressure transducer

DATE INVENTORIED (C711) month – day – year

RECORD READY FOR WEB (C32): Y ready to display, C condi-tional, P proprie-tary, L local use only

REMARKS (C806)

FOOTNOTES

1 SITE TYPE (C802)

GL	Glacier	OC	Ocean	GW	Well	SB	Subsurface
WE	Wetland	OC-CO	Coastal	GW-CR	Collector or Ranney type well	SB-CV	Cave
AT	Atmosphere	LK	Lake, Reservoir, Impoundment	GW-EX	Extensometer well	SB-GWD	Groundwater drain
ES	Estuary			GW-HZ	Hyporheic-zone well	SB-TSM	Tunnel, shaft, or mine
LA	Land	SP	Spring	GW-IW	Interconnected wells	SB-UZ	Unsaturated zone
LA-EX	Excavation	ST	Stream	GW-TH	Test hole not completed as a well		
LA-OU	Outcrop	ST-CA	Canal	GW-MW	Multiple wells		
LA-SNK	Sinkhole	ST-DCH	Ditch				
LA-SH	Soil hole	ST-TS	Tidal stream				
LA-SR	Shore	FA-WIW	Waste-Injection well				

2 WS water supply, DO domestic, CO commer-cial, IN industrial, IR irrigation, MI mining, LV livestock, PH power hydro-electric, ST waste water treatment, RM remedia-tion, TE thermo-electric power, AQ aqua-culture

C22 Other (see manual for codes)
C36 Other (see manual for codes)
C39 is mandatory for all sites having data in SWUDS.

Figu 3 Groundw e Schedule, o 9 4 A

GENERAL SITE DATA

DATA RELIAB LITY (C3): C field checked | L poor location | M minimal data | U unchecked

DATE OF FIRST CONSTRUCTION (C21): month – day – year

USE OF SITE (C23): A anode | C standby emer. supply | D drain | E geothermal | G seismic | H heat reservoir | M mine | O observation | P oil or gas | R recharge | S repressurize | T test | U unused | V withdrawal/return | W withdrawal | X waste | Z destroyed

SECONDARY USE OF SITE (C301) (See use of site)

TERTIARY USE OF SITE (C302) (See use of site)

USE OF WATER (C24): A air cond. | B bottling | C commercial | D dewater | E power | F fire | H domestic | I irrigation | J industrial (cooling) | K mining | M medicinal | N industrial | P public supply | Q aquaculture | R recreations | S stock | T institutional | U unused | Y desalination | Z other

SECONDARY USE OF WATER (C25) (see use of water)

TERTIARY USE OF WATER (C26) (see use of water)

AQUIFER TYPE (C713): U unconfined single | N unconfined multiple | C confined single | M confined multiple | X mixed

PRIMARY AQUIFER (C714)

NATIONAL AQUIFER (C715)

HOLE DEPTH (C27) .

WELL DEPTH (C28) .

SOURCE OF DEPTH DATA (C29): A other gov't | D driller | G geologist | L logs | M memory | O owner | R other reported | S reporting agency | Z other

WATER-LEVEL DATA

DATE WATER-LEVEL MEASURED (C235): month – day – year

TIME (C709)

WATER-LEVEL TYPE CODE (C243): L land surface | M meas. pt. | S vertical datum

WATER LEVEL (C237/241/242) .

MP SEQUENCE NO. (C248) (Mandatory if WL type=M)

WATER-LEVEL DATUM (C245) (Mandatory if WL type=S): NGVD29 National Geodetic Vertical Datum Of 1929 | NAVD88 North American Vertical Datum Of 1988 | Other (See manual for codes)

SITE STATUS FOR WATER LEVEL (C238): A atmos. pressure | B tide stage | C ice | D dry | E recently flowing | F flowing | G nearby flowing | H nearby recently flowing | I injector site | J injector site monitor | M plugged | N measurement discontinued | O obstruction | P pumping | R recently pumped | S nearby pumping | T nearby recently pumped | V foreign substance | W well destroyed | X affected by surface water | Z other

METHOD OF WATER-LEVEL MEASUREMENT(C239): A airline | B analog | C calibrated airline | D differential GPS | E estimated | F transducer | G pressure gage | H calibrated press. gage | L geophysical logs | M manometer | N non-rec. gage | O observed | P acoustic pulse | R reported | S steel tape | T electric tape | V calibrated elec. tape | Z other

WATER-LEVEL ACCURACY (C276): 0 foot | 1 tenth | 2 hundredth | 9 not to nearest foot

SOURCE OF WATER-LEVEL DATA (C244): A other gov't | D driller's log | G geologist | L geophysical logs | M memory | O owner | R other reported | S reporting agency | Z other

PERSON MAKING MEASUREMENT (C246) (WATER LEVEL PARTY)

MEASUR NG AGENCY (C247) (SOURCE)

EQUIP ID (C249) (20 char) ______

REMARKS (C267) (256 char) ______

RECORD READY FOR WEB (C858): Y ready to display | C conditional | P proprietary | L local use only

CONSTRUCTION DATA

RECORD TYPE (C754): CONS

RECORD SEQUENCE NO. (C723)

DATE OF COMPLETED CONSTRUCTION (C60): month – day – year

NAME OF CONTRACTOR (C63)

SOURCE OF DATA (C64): A other gov't | D driller | G geologist | L logs | M memory | O owner | R other reported | S reporting agency | Z other

METHOD OF CONSTRUCTION (C65): A air-rotary | B bored or augered | C cable tool | D dug | H hydraulic rotary | J jetted | P air percussion | R reverse rotary | S sonic | T trenching | V driven | W drive wash | Z other

TYPE OF FINISH (C66): C porous concrete | F gravel w/perf. | G gravel screen | H horiz. gallery | O open end | P perf or slotted | S screen | T sand point | W walled | X open hole | Z other

TYPE OF SEAL (C67): B bentonite | C clay | G cement grout | N none | Z other

BOTTOM OF SEAL (C68)

METHOD OF DEVELOPMENT (C69): A air-lift pump | B bailed | C compressed air | J jetted | N none | P pumped | S surged | Z other

HOURS OF DEVELOPMENT (C70)

SPECIAL TREATMENT (C71): C chemicals | D dry ice | E explosives | F deflocculent | H hydrofracturing | M mechanical | Z other

2 - Groundwater Site Schedule

CONSTRUCTION HOLE DATA (3 sets shown)

RECORD TYPE (C756) HOLE   RECORD SEQUENCE NO. (C724)   SEQUENCE NO. OF PARENT RECORD (C59)

DEPTH TO TOP OF INTERVAL (C73)   DEPTH TO BOTTOM OF NTERVAL (C74)   DIAMETER OF INTERVAL (C75)

RECORD SEQUENCE NO. (C724)

DEPTH TO TOP OF INTERVAL (C73)   DEPTH TO BOTTOM OF NTERVAL (C74)   DIAMETER OF INTERVAL (C75)

RECORD SEQUENCE NO. (C724)

DEPTH TO TOP OF INTERVAL (C73)   DEPTH TO BOTTOM OF NTERVAL (C74)   DIAMETER OF INTERVAL (C75)

CONSTRUCTION CASING DATA (4 sets shown)

RECORD TYPE (C758) CSNG   RECORD SEQUENCE NO. (C725)   SEQUENCE NO. OF PARENT RECORD (C59)

DEPTH TO TOP OF CAS NG (C77)   DEPTH TO BOTTOM OF CASING (C78)   DIAMETER OF CASING (C79)

4 CASING MATERIAL (C80)   CASING THICKNESS (C81)

RECORD SEQUENCE NO. (C725)   SEQUENCE NO. OF PARENT RECORD (C59)

DEPTH TO TOP OF CASING (C77)   DEPTH TO BOTTOM OF CASING (C78)   DIAMETER OF CASING (C79)

4 CASING MATERIAL (C80)   CASING THICKNESS (C81)

RECORD SEQUENCE NO. (C725)   SEQUENCE NO. OF PARENT RECORD (C59)

DEPTH TO TOP OF CAS NG (C77)   DEPTH TO BOTTOM OF CASING (C78)   DIAMETER OF CASING (C79)

4 CAS NG MATERIAL (C80)   CAS NG THICKNESS (C81)

RECORD SEQUENCE NO. (C725)   SEQUENCE NO. OF PARENT RECORD (C59)

DEPTH TO TOP OF CAS NG (C77)   DEPTH TO BOTTOM OF CASING (C78)   DIAMETER OF CASING (C79)

4 CAS NG MATERIAL (C80)   CAS NG THICKNESS (C81)

FOOTNOTE:

4 CAS NG MATERIAL CODES

A	B	C	D	E	F	G	H	I	J	K	L	M	N	P	Q	R	S	T	U	V	W	X	Y	Z	4	6
abs	brick	concrete	copper	PTFE	Fiber-glass	galv. iron	Fiber-glass plastic	wrought iron	Fiber-glass epoxy	PVC thread-ed	glass	other metal	PVC glued	PVC or plastic	FEP	rock or stone	steel	tile	coated steel	stain-less steel	wood	steel carbon	steel galva-nized	other mat.	stain-less 304	stain-less 316

CONSTRUCTION OPENINGS DATA (3 sets shown)

RECORD TYPE (C760) O P E N

RECORD SEQUENCE NO. (C726)

SEQUENCE NO. OF PARENT RECORD (C59)

DEPTH TO TOP OF INTERVAL (C83)

DEPTH TO BOTTOM OF INTERVAL (C84)

DIAMETER OF INTERVAL (C87)

[5] MATERIAL TYPE (C86)

[6] TYPE OF OPENING (C85)

LENGTH OF OPENING (C89)

WIDTH OF OPENING (C88)

RECORD SEQUENCE NO. (C726)

DEPTH TO TOP OF INTERVAL (C83)

DEPTH TO BOTTOM OF INTERVAL (C84)

DIAMETER OF INTERVAL (C87)

[5] MATERIAL TYPE (C86)

[6] TYPE OF OPENING (C85)

LENGTH OF OPENING (C89)

WIDTH OF OPENING (C88)

RECORD SEQUENCE NO. (C726)

DEPTH TO TOP OF INTERVAL (C83)

DEPTH TO BOTTOM OF INTERVAL (C84)

DIAMETER OF INTERVAL (C87)

[5] MATERIAL TYPE (C86)

[6] TYPE OF OPENING (C85)

LENGTH OF OPENING (C89)

WIDTH OF OPENING (C88)

FOOTNOTES:

[5] TYPE OF MATERIAL CODES FOR OPEN SECTIONS

A	B	C	D	E	F	G	H	I	J	K	L	M	N	P	Q	R	S	T	V	W	X	Y	Z	4	6
ABS	brass or bronze	concrete	ceramic	PTFE	fiber-glass	galv. iron	fiber-glass plastic	wrought iron	fiber-glass epoxy	PVC thread-ed	glass	other metal	PVC glued	PVC	FEP	stain-less steel	steel	tile	brick	mem-brane	steel carbon	steel galva-nized	other	stain-less 304	stain-less 316

[6] TYPE OF OPENINGS CODES

F	L	M	P	R	S	T	W	X	Z
fractured rock	louvered or shutter-type	mesh screen	perforated, porous or slotted	wire-wound screen	screen (unk.)	sand point screen	walled or shored	open hole	other

CONSTRUCTION MEASURING POINT DATA

RECORD TYPE (C766) M P N T

RECORD SEQUENCE NO. (C728)

BEGINNING DATE (C321) month – day – year

ENDING DATE (C322)

M.P. HEIGHT (C323)

ALTITUDE OF MEASURING POINT (C325)

ALTITUDE METHOD (C326)

ALTITUDE ACCURACY (C327)

ALTITUDE DATUM (C328)

M.P. REMARKS (C324)

RECORD READY FOR WEB (C857)

Y	C	P	L
ready to display	condi-tional	proprie-tary	local use only

CONSTRUCTION LIFT DATA

RECORD TYPE (C752) L I F T

RECORD SEQUENCE NO. (C254)

TYPE OF LIFT (C43): A air, B bucket, C centri-fugal, J jet, P piston, R rotary, S submer-sible, T turbine, U un-known, X no lift, Z other

DATE RECORDED (C38) month – day – year

PUMP INTAKE DEPTH (C44)

TYPE OF POWER (C45): D diesel, E electric, G gaso-line, H hand, L LP gas, N natural gas, S solar, W windmill, Z other

HORSE-POWER RATING (C46) .

MANUFACTURER (C48)

SERIAL NO. (C49)

POWER COMPANY (C50)

POWER COMPANY ACCOUNT NUMBER (C51)

POWER METER NUMBER (C52)

PUMP RATING (C53) (million gallons/units of fuel) .

ADDITIONAL LIFT (C255)

PERSON OR COMPANY MAINTAINING PUMP (C54)

RATED PUMP CAPACITY (gpm) (C268)

STANDBY POWER (C56) (see TYPE OF POWER)

HORSEPOWER OF STANDBY POWER SOURCE (C57) .

MISCELLANEOUS OWNER DATA

RECORD TYPE (C768) O W N R

RECORD SEQUENCE NO. (C718)

DATE OF OWNERSHIP (C159) – –

WU OWNER TYPE (C350): CP Corporation, GV Govern-ment, IN Individual, MI Military, OT Other, TG Tribal, WS Water Supplier

END DATE OF OWNERSHIP (C374) – –

OWNER'S NAME (C161)

EXAMPLES: JONES, RALPH A.
JONES CONSTRUCTION COMPANY

OWNER'S PHONE NUMBER (C351)

ACCESS TO OWNER'S NAME (C352): 0 Public Access, 1 Coop-erator, 2 USGS Only, 3 District Only, 4 Proprietary

OWNER'S ADDRESS (LINE 1) (C353)

OWNER'S ADDRESS (LINE 2) (C354)

OWNER'S CITY NAME (C355)

STATE (C356)

OWNER'S ZIP CODE (C357) –

OWNER'S COUNTRY NAME (C358)

ACCESS TO OWNER'S PHONE/ADDRESS (C359): 0 Public Access, 1 Coop-erator, 2 USGS Only, 3 District Only, 4 Proprietary

MISCELLANEOUS VISIT DATA

RECORD TYPE (C774) V I S T

RECORD SEQUENCE NO. (C737)

DATE OF VISIT (C187) month – day – year

NAME OF PERSON (C188)

MISCELLANEOUS OTHER ID DATA (2 sets shown)

RECORD TYPE (C770) O T I D

RECORD SEQUENCE NO. (C736) | OTHER ID (C190)

ASSIGNER (C191)

RECORD SEQUENCE NO. (C736) | OTHER ID (C190)

ASSIGNER (C191)

MISCELLANEOUS OTHER DATA

RECORD TYPE (C772) O T D T | RECORD SEQUENCE NO. (C312)

OTHER DATA TYPE (C181)

OTHER DATA LOCATION (C182) C Cooperator's Office, D District Office R Reporting Agency Z other

DATA FORMAT (C261) F files, M machine readable, P published, Z other

MISCELLANEOUS LOGS DATA (3 sets shown)

RECORD TYPE (C778) L O G S | RECORD SEQUENCE NO. (C739) | TYPE OF LOG (C199)

BEG NN NG DEPTH (C200) . | ENDING DEPTH (C201) . | SOURCE OF DATA (C202) A other gov't D driller G geol-ogist L logs M memory O owner R other reported S reporting agency Z other

DATA FORMAT (C225) F files M machine readable P published Z other | OTHER DATA LOCATION (C226)

RECORD TYPE (C778) L O G S | RECORD SEQUENCE NO. (C739) | TYPE OF LOG (C199)

BEG NNING DEPTH (C200) . | ENDING DEPTH (C201) . | SOURCE OF DATA (C202) A other gov't D driller G geol-ogist L logs M memory O owner R other reported S reporting agency Z other

DATA FORMAT (C225) F files M machine readable P published Z other | OTHER DATA LOCATION (C226)

RECORD TYPE (C778) L O G S | RECORD SEQUENCE NO. (C739) | TYPE OF LOG (C199)

BEGINNING DEPTH (C200) . | ENDING DEPTH (C201) . | SOURCE OF DATA (C202) A other gov't D driller G geol-ogist L logs M memory O owner R other reported S reporting agency Z other

DATA FORMAT (C225) F files M machine readable P published Z other | OTHER DATA LOCATION (C226)

ACOUSTIC LOG:
AS Sonic
AV Acoustic velocity
AW Acoustic waveform
AT Acous ic televiewer

CALIPER LOG:
CP Caliper
CS Caliper, single arm
CT Caliper, three arm
CM Caliper, multi arm
CA Caliper, acous ic

DRILLING LOG:
DT Drilling ime
DR Drillers
DG Geologists
DC Core

ELECTRIC LOG:
EE Electric
ER Single-point resistance
EP Spontaneous potential
EL Long-normal resistivity
ES Short-normal resis ivity
EF Focused resistivity
ET Lateral resistivity
EN Microresistivity
EC Microresistivity, forused
EO Microresis ivity, lateral
ED Dipmeter

ELECTROMAGNETIC LOG:
MM Magnetic log
MS Magnetic susceptibiity log
MI Electromagne ic induc ion log
MD Electromagne ic dual induction log
MR Radar reflection image log
MV Radar direct-wave velocity log
MA Radar direct-wave amplitude log

FLUID LOG:
FC Fluid conductivity
FR Fluid resistivity
FT Fluid temperature
FF Fluid differen ial temperature
FV Fluid velocity
FS Spinner flowmeter
FH Heat-pulse flowmeter
FE Electromagnetic flowmeter
FD Doppler flowmeter
FA Radioactive tracer
FY Dye tracer
FB Brine tracer

NUCLEAR LOG:
NG Gamma
NS Spectral gamma
NA Gamma-gamma
NN Neutron
NT Neutron activitation
NM Neuclear magnetic resonance

OPTICAL LOG:
OV Video
OF Fisheye video
OS Sidewall video
OT Optical televiewer

COMBINATION LOG:
ZF Gamma, fluid resistivity, temperature
ZI Gamma, electromagne ic induction
ZR Long/short normal resistivity
ZT Fluid resis ivity, temperature
ZM Electromagnetic flowmeter, fluid resistivity, temperature
ZN Long/short normal resistivity, spontaneous poten ial
ZP Single-point resistance, spontaneous potential
ZE Gamma, long/short normal resistivity, spontaneous potential, single-point resistance, fluid resi ivity, temperature

WELL CONSTRUCTION LOG:
WC Casing collar
WD Borehold deviation

OTHER LOG:
OR Other

MISCELLANEOUS NETWORK DATA (3 types shown)

RECORD TYPE (C780) N E T W RECORD SEQUENCE NO. (C730) TYPE OF NETWORK (C706) Q W water quality BEGINNING YEAR (C115) ENDING YEAR (C116)

TYPE OF ANALYSIS (C120)

A	B	C	D	E	F	G	H	I	J	K	L	M	N	P	Z
physical proper-ties	common ions	trace elements	pesti-cides	nutri-ents	sanitary analysis	codes D&B	codes B&E	codes B&C	codes B&F	codes D&E	codes C,D&E	all or most	codes B&C& radio-active	codes B,C&A	other

SOURCE AGENCY (C117) [7] FREQUENCY OF COLLECTION (C118) ANALYZING AGENCY (C307) [8] PRIMARY NETWORK SITE (C257) [8] SECONDARY NETWORK SITE (C708)

RECORD TYPE (C780) N E T W RECORD SEQUENCE NO. (C730) TYPE OF NETWORK (C706) W L water level BEGINNING YEAR (C115) ENDING YEAR (C116)

SOURCE AGENCY (C117) [7] FREQUENCY OF COLLECTION (C118) [8] PRIMARY NETWORK SITE (C257) [8] SECONDARY NETWORK SITE (C708)

RECORD TYPE (C780) N E T W RECORD SEQUENCE NO. (C730) TYPE OF NETWORK (C706) W D pumpage or with-drawals BEGINNING YEAR (C115) ENDING YEAR (C116)

SOURCE AGENCY (C117) [7] FREQUENCY OF COLLECTION (C118) METHOD OF COLLECTION (C133)

C	E	M	U	Z
calcu-lated	esti-mated	meter-ed	un-known	other

[8] PRIMARY NETWORK SITE (C257) [8] SECONDARY NETWORK SITE (C708)

FOOTNOTES:

[7] FREQUENCY OF COLLECTION CODES

A	B	C	D	F	I	M	O	Q	S	W	Z	2	3	4	5	X
annually	bi monthly	continu-ously	daily	semi-monthly	inter mittent	monthly	one-time only	quarter-ly	semi-annually	weekly	other	bi-annually	every 3 years	every 4 years	every 5 years	every 10 years

[8] NETWORK SITE CODES

1	2	3	4
national,	district,	project,	co-operator.

MISCELLANEOUS REMARKS DATA (4 types shown)

RECORD TYPE (C788) R M K S RECORD SEQUENCE NO. (C311) DATE OF REMARK (C184) month – day – year

REMARKS (C185)

Subsequent entries may be used to continue the remark. Miscellaneous remarks field is limited to 256 characters.

RECORD TYPE (C788) R M K S RECORD SEQUENCE NO. (C311) DATE OF REMARK (C184) month – day – year

REMARKS (C185)

Subsequent entries may be used to continue the remark. Miscellaneous remarks field is limited to 256 characters.

DISCHARGE DATA

RECORD SEQUENCE NO. (C147)

DATE DISCHARGE MEASURED (C148) — month — day — year

TYPE OF DISCHARGE (C703): P (pumped), F (flow)

DISCHARGE (gpm) (C150)

ACCURACY OF DISCHARGE MEASUREMENT (C310): E (excellent (LT 2%)), G (good (2%-5%)), F (fair (5%-8%)), P (poor (GT 8%))

SOURCE OF DATA (C151): A (other gov't), D (driller), G (geologist), L (logs), M (memory), O (owner), R (other reported), S (reporting agency), Z (other)

METHOD OF DISCHARGE MEASUREMENT (C152): A (acoustic meter), B (bailer), C (current meter), D (Doppler meter), E (estimated), F (flume), M (totaling meter), O (orifice), P (pitot-tube), R (reported), T (trajectory), U (venturi meter), V (volumetric meas), W (weir), X (unknown), Z (other)

PRODUCTION WATER LEVEL (C153)

STATIC WATER LEVEL (C154)

SOURCE OF DATA (C155): A (other gov't), D (driller), G (geologist), L (logs), M (memory), O (owner), R (other reported), S (reporting agency), Z (other)

METHOD OF WATER LEVEL MEASUREMENT (C156): A (airline), B (recorder), C (calibrated airline), E (estimated), G (pressure gage), H (calibrated press. gage), L (geophysical logs), M (manometer), N (non-rec. gage), R (reported), S (steel tape), T (electric tape), U (unknown), V (calibrated elec. tape), Z (other)

PUMPING PERIOD (C157)

SPECIFIC CAPACITY (C272)

DRAWDOWN (C309)

GEOHYDROLOGIC DATA

RECORD TYPE (C748): G E O H

RECORD SEQUENCE NO. (C721)

DEPTH TO TOP OF UNIT (C91)

DEPTH TO BOTTOM OF UNIT (C92)

UNIT IDENTIFIER (C93)

LITHOLOGY (C96)

CONTRIBUTING UNIT (C304): P (principal aquifer), S (secondary aquifer), N (no contribution), U (unknown)

LITHOLOGIC MODIFIER (C97)

GEOHYDROLOGIC AQUIFER DATA

RECORD TYPE (C750): A Q F R

RECORD SEQUENCE NO. (C742)

SEQUENCE NO. OF PARENT RECORD (C256)

DATE (C95) — month — day — year

STATIC WATER LEVEL (C126)

CONTRIBUTION (C132)

SITE LOCATION SKETCH AND DIRECTIONS

Township ___________ Range ___________

Section # ___________

8 - Ground-water site schedule

References

Bureau of Reclamation, 1967, Water measurement manual, A water resources technical publication: Washington, D.C., U.S. Government Printing Office, p. 199.

Driscoll, F.G., 1966, Groundwater and wells: St. Paul, Minnesota, Johnson Filtration Systems, Inc., 440 p.

Hoopes, B.C., ed., 2004, User's manual for the National Water Information System of the U.S. Geological Survey, Ground-Water Site-Inventory System (version 4.4): U.S. Geological Survey Open-File Report 2005–1251, 274 p.

Lawrence, F.E., and Braunworth, P.L., 1906, Fountain flow of water in vertical pipes: Transactions of the American Society of Civil Engineers, v. 57, p. 265–306.

GWPD 8—Estimating discharge from a pumped well by use of the trajectory free-fall or jet-flow method

VERSION: 2010.1

PURPOSE: To estimate the discharge from a pumped well from a non-vertical standard pipe by using the trajectory free-fall or jet-flow method.

Materials and Instruments

1. L-shaped measuring device (carpenter's square)
2. Support for measuring device
3. Small hand level
4. Clamp
5. Field notebook
6. Pencil or pen, blue or black ink. Strikethrough, date, and initial errors; no erasures
7. Groundwater Site Inventory (GWSI) System Ground-water Site Schedule, Form 9-1904-A

Data Accuracy and Limitations

1. Under ordinary field conditions, with reasonable care, measurements can be made in which the error seldom exceeds 10 percent.
2. The most accurate estimated discharge will be obtained when the pipe is truly horizontal.
3. The discharge pipe should be a straight length of standard pipe at least 5 feet long, so that the open end is at least this distance from the nearest elbow or bend in the pipe.
4. If the discharge pipe slopes upward, the estimated discharge will be too high; if it slopes downward, the estimated discharge will be too low.
5. The principal difficulty with using this method is in measuring the coordinates (X and Y) of the jet-flow stream accurately.
6. Well flow should be constant so that the top of the stream at the open end of the pipe does not vary appreciably.
7. Not accurate for small flows. For small flows, measure the well discharge with a flowmeter or a bucket and stopwatch.

Advantages

1. This method provides a simple, inexpensive, and practical means of estimating flow from horizontal and inclined pipes for field tests.
2. No special training is needed to use this method.

Disadvantages

1. This method provides only an approximate discharge from wells with horizontal or inclined pipes.
2. Well flow should be constant. The top of the stream at the open end of the pipe should not vary appreciably.

Assumptions

1. The discharge pipe does not have a circular orifice weir.
2. The discharge pipe does not have an in-line flowmeter.

Instructions

1. Measure the inside diameter (D) of the pipe accurately, in inches (fig. 1*A*).
2. Measure the distance (X) that the jet flow of water travels, in inches parallel to the top of the pipe for a 12-inch vertical drop (Y; fig. 1*B*).
3. If the jet flow is brooming or spreading from the end of the horizontal pipe, the center of the falling stream (P) can be located more reliably than can a point on the surface of the stream. When brooming or spreading flow occurs, measure X from the center of the pipe for a 12-inch vertical drop, and measure Y from the center of the pipe to the center of the falling stream (fig. 1*C*).
4. Estimate well discharge by using the discharge curves for measurement of flow from non-vertical standard pipes (fig. 2). For example, see the sample calculation in figure 2 for a 5-inch well with a jet stream of 16 inches (X) and a 12-inch vertical drop (Y). Discharge from this well is about 330 gallons per minute.
5. For partially filled non-vertical pipes, measure the freeboard (F) and the inside diameter (D) of the pipe (fig. 1C). Calculate the ratio of F/D as a percentage. Measure the distance X of the jet stream for a 12-inch vertical drop (Y), and estimate a well discharge using the discharge curves in figure 2. The actual estimated discharge will be the value for a full pipe multiplied by a correction factor obtained from table 1. Use the correction factor in the column opposite the ratio of F/D calculated above for the partially filled non-vertical pipe.
6. Record estimated discharge in the field notebook and in the discharge data section on the GWSI Groundwater Site Schedule (fig. 3, Form 9-1904-A).

Data Recording

Data are recorded in a field notebook. Discharge data should also be recorded in the discharge data section of the GWSI Groundwater Site Schedule (Form 9-1904-A). This is best described as a trajectory method and should be coded as "T" in field C152 on Form 9-1904-A.

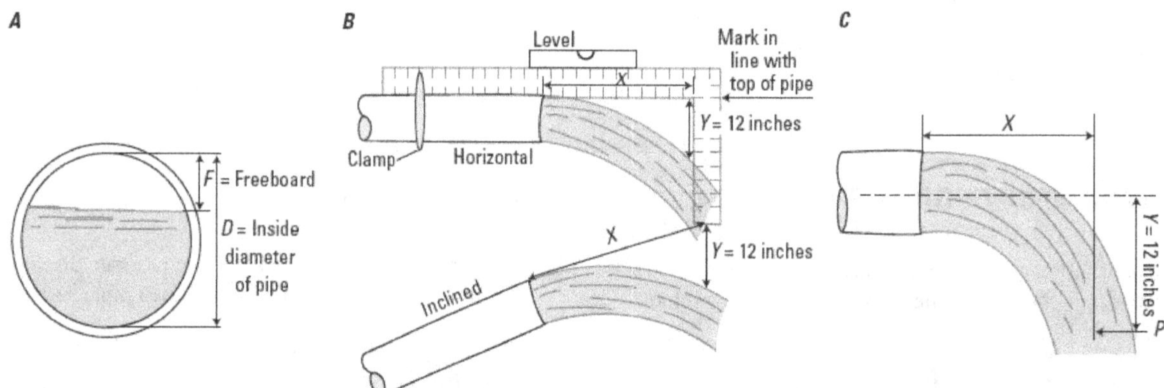

Figure 1. Measurements for estimating flow from (*A*) a partially filled pipe (Anderson, 1963), (*B*) a horizontal or inclined pipe with steady flow (Anderson, 1963), and (*C*) a horizontal pipe when brooming or spreading flow occurs (Driscoll, 1986).

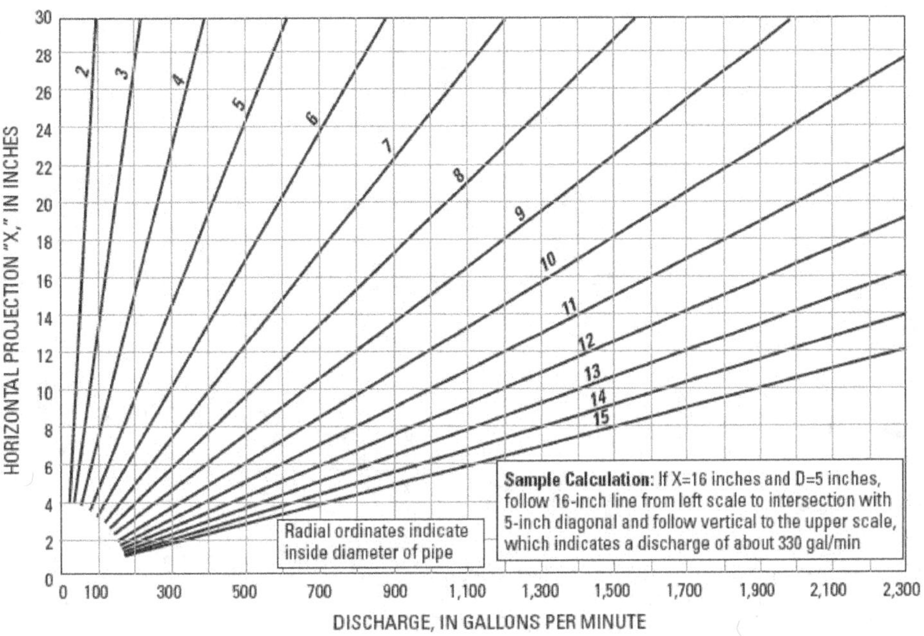

Figure 2. Discharge curves for measurement of flow from non-vertical standard pipes based on a constant value of 12 inches for *Y*. If the discharge in the pipe is not flowing full, multiply the discharge by the correction factor found in table 1 (McDonald, 1950).

Table 1. Correction factors for percentages of discharge (see fig. 2).

[F, freeboard; D, inside diameter]

F/D percent	Correction factor
5	0.981
10	.948
15	.905
20	.858
25	.805
30	.747
35	.688
40	.627
45	.564
50	.500
55	.436
60	.375
65	.312
70	.253
75	.195
80	.142
85	.095
90	.052
95	.019
100	.000

References

Anderson, K.E., 1963, Water well handbook (2d ed.): Missouri Water Well Drillers Association, p. 156.

Bureau of Reclamation, 1975, Water measurement manual, A water resources technical publication (2d ed., reprinted): U.S. Department of the Interior, p. 200.

Driscoll, F.G., 1986, Groundwater and wells (2d ed.): St. Paul, Minnesota, Johnson Filtration Systems, Inc., 1089 p.

Hoopes, B.C., ed., 2004, User's manual for the National Water Information System of the U.S. Geological Survey, Ground-Water Site-Inventory System (version 4.4): U.S. Geological Survey Open-File Report 2005–1251, 274 p.

McDonald, H.R., 1950, How to estimate flow from pipes: Engineering News-Record, August 31, 1950, p. 48.

FORM NO. 9-1904-A
Revised Sept 2009, NWIS 4.9

File Code ______
Date ______

Coded by ______
Checked by ______
Entered by ______

U.S DEPT. OF THE INTERIOR
GEOLOGICAL SURVEY

GROUNDWATER SITE SCHEDULE
General Site Data

AGENCY CODE (C4) USGS | SITE ID (C1) | PROJECT (C5)

STATION NAME (C12/900)

SITE TYPE[1] (C802) Primary - Secondary | DISTRICT (C6) | COUNTRY (C41) | STATE (C7)

COUNTY or TOWN (C8) | County code

LATITUDE (C9) . | LONGITUDE (C10) . | LAT/LONG ACCURACY (C11): H (Hndrth sec.) 1 (tenth sec.) 5 (half sec.) S (sec.) R (3 sec.) F (5 sec.) T (10 sec.) M (min.) U (Un-known)

LAT/LONG METHOD (C35): C (land net) D (DGPS) G (GPS) L (LORAN) M (map) N (inter-polated digital map) R (reported) S (survey) U (un-known)

LAT/LONG DATUM (C36): NAD27 (North American Datum of 1927) NAD83 (North American Datum of 1983)

ALTITUDE (C16) .

ALTITUDE ACCURACY (C18)

ALTITUDE METHOD (C17): A (altimeter) D (DGPS) G (GPS) I (IfSAR) J (LIDAR) L (Level) M (map) N (DEM) R (re-ported) U (un-known)

ALTITUDE DATUM (C22): NGVD29 (National Geodetic Vertical Datum of 1929) NAVD88 (North American Vertical Datum of 1988)

LAND NET (C13): ¼ ¼ ¼ S section T township range merid

TOPOGRAPHIC SETTING (C19): A (alluvial fan) B (playa) C (stream channel) D (depres-sion) E (dunes) F (flat) G (flood-plain) H (hill-top) K (sink-hole) L (lake or swamp) M (mangrove swamp) O (off-shore) P (pedi-ment) S (hill-side) T (ter-race) U (undu-lating) V (valley flat) W (upland draw)

HYDROLOGIC UNIT CODE (C20) | DRAINAGE BASIN CODE (C801) | STANDARD TIME ZONE (C813) | DAYLIGHT SAVINGS TIME FLAG (C814) **Y** OR **N**

MAP NAME (C14) | MAP SCALE (C15)

AGENCY USE (C803): A (active no/na) D (discon-tinued) I (inactive site) L (active written) M (active oral) O (inventory site) R (remediated)

[2] NATIONAL WATER-USE (C39)

DATA TYPE (C804)
Place an 'A' (active), an 'I' (inactive), or an 'O' (inventory) in the appropriate box
WL cont, WL int, QW cont, QW int, PR cont, PR int, EV cont, EV int, wind vel., tide cont, tide int, sed. con, sed. ps, peak flow, low flow, state water use

INSTRUMENTS (C805)
(Place a "Y" in the appropriate box):
digital rec-order, graphic rec-order, tele-metry land line, tele-metry radio, tele-metry satellite, AHDAS, crest-stage gage, tide gage, deflec-tion meter, bubble gage, stilling well, CR type recorder, weigh-ing rain gage, tipping bucket rain gage, acoustic velocity meter, electro-magnetic flowmeter, pressure transducer

DATE INVENTORIED (C711) month - day - year

RECORD READY FOR WEB (C32): Y (ready to display) C (condi-tional) P (proprie-tary) L (local use only)

REMARKS (C806)

FOOTNOTES

1SITE TYPE (C802)

GL	Glacier	OC	Ocean	GW	Well	SB	Subsurface
WE	Wetland	OC-CO	Coastal	GW-CR	Collector or Ranney type well	SB-CV	Cave
AT	Atmosphere	LK	Lake, Reservoir, Impoundment	GW-EX	Extensometer well	SB-GWD	Groundwater drain
ES	Estuary			GW-HZ	Hyporheic -zone well	SB-TSM	Tunnel, shaft, or mine
LA	Land	SP	Spring	GW-IW	Interconnected wells	SB-UZ	Unsaturated zone
LA-EX	Excavation	ST	Stream	GW-TH	Test hole not completed as a well		
LA-OU	Outcrop	ST-CA	Canal	GW-MW	Multiple wells		
LA-SNK	Sinkhole	ST-DCH	Ditch				
LA-SH	Soil hole	ST-TS	Tidal stream				
LA-SR	Shore	FA-WIW	Waste-Injection well				

[2] WS (water supply) DO (domestic) CO (commer-cial) IN (industrial) IR (irrigation) MI (mining) LV (livestock) PH (power hydro-electric) ST (waste water treatment) RM (remedia-tion) TE (thermo-electric power) AQ (aqua-culture)

C22 Other (see manual for codes)
C36 Other (see manual for codes)
C39 is mandatory for all sites having data in SWUDS.

Figure 3. Groundwater Site Schedule, Form 9-1904-A.

GENERAL SITE DATA

DATA RELIAB LITY (C3) C L M U — field checked, poor location, minimal data, un-checked

DATE OF FIRST CONSTRUCTION (C21) month – day – year

USE OF SITE (C23) A anode, C standby emer. supply, D drain, E geo-thermal, G seismic, H heat reservoir, M mine, O obser-vation, P oil or gas, R recharge, S repres-surize, T test, U unused, V with-drawal/ return, W with-drawal, X waste, Z des-troyed

SECOND-ARY USE OF SITE (C301) (See use of site)

TERTIARY USE OF SITE (C302) (See use of site)

USE OF WATER (C24) A air cond., B bottling, C comm-ercial, D de-water, E power, F fire, H domes-tic, I irri-gation, J indus-trial (cooling), K mining, M medi-cinal, N indus-trial, P public supply, Q aqua-culture, R recrea-tions, S stock, T insti-tutional, U unused, Y desalin-ation, Z other

SECOND-ARY USE OF WATER (C25) (see use of water)

TERTIARY USE OF WATER (C26) (see use of water)

AQUIFER TYPE (C713) U unconfined single, N unconfined multiple, C confined single, M confined multiple, X mixed

PR MARY AQU FER (C714)

NATIONAL AQU FER (C715)

HOLE DEPTH (C27)

WELL DEPTH (C28)

SOURCE OF DEPTH DATA (C29) A other gov't, D driller, G geol-ogist, L logs, M memory, O owner, R other reported, S reporting agency, Z other

WATER-LEVEL DATA

DATE WATER-LEVEL MEASURED (C235) month – day – year

T ME (C709)

WATER-LEVEL TYPE CODE (C243) L M S — land surface, meas. pt., vertical datum

WATER LEVEL (C237/241/242)

MP SEQUENCE NO. (C248) (Mandatory if WL type=M)

WATER-LEVEL DATUM (C245) (Mandatory if WL type=S) NGVD29 National Geodetic Vertical Datum Of 1929, NAVD88 North American Vertical Datum Of 1988, Other (See manual for codes)

SITE STATUS FOR WATER LEVEL (C238) A atmos. pressure, B tide stage, C ice, D dry, E recently flowing, F flowing, G nearby flowing, H nearby recently flowing, I injector site, J injector site monitor, M plugged, N measure-ment discontinued, O obstruc-tion, P pumping, R recently pumped, S nearby pumping, T nearby recently pumped, V foreign sub-stance, W well des-troyed, X affected by surface water, Z other

METHOD OF WATER-LEVEL MEASUREMENT(C239) A airline, B analog, C calibrated airline, D differ-ential GPS, E esti-mated, F trans-ducer, G pressure gage, H calibrated press. gage, L geophysi-cal logs, M mano-meter, N non-rec. gage, O observed, P acoustic pulse, R reported, S steel tape, T electric tape, V calibrated elec. tape, Z other

WATER-LEVEL ACCURACY (C276) 0 foot, 1 tenth, 2 hun-dredth, 9 not to nearest foot

SOURCE OF WATER-LEVEL DATA (C244) A other gov't, D driller's log, G geol-ogist, L geophysi-cal logs, M memory, O owner, R other reported, S reporting agency, Z other

PERSON MAK NG MEASUREMENT (C246) (WATER LEVEL PARTY)

MEASURING AGENCY (C247) (SOURCE)

EQU P D (C249) (20 char)

REMARKS (C267) (256 char)

RECORD READY FOR WEB (C858) Y ready to display, C condi-tional, P proprie-tary, L local use only

CONSTRUCTION DATA

RECORD TYPE (C754) CONS

RECORD SEQUENCE NO. (C723)

DATE OF COMPLETED CONSTRUCTION (C60) month – day – year

NAME OF CONTRACTOR (C63)

SOURCE OF DATA (C64) A other gov't, D driller, G geol-ogist, L logs, M memory, O owner, R other reported, S reporting agency, Z other

METHOD OF CONSTRUCTION (C65) A air-rotary, B bored or augered, C cable tool, D dug, H hydraulic rotary, J jetted, P air per-cussion, R reverse rotary, S sonic, T trenching, V driven, W drive wash, Z other

TYPE OF F NISH (C66) C porous concrete, F gravel w/perf., G gravel screen, H horiz. gallery, O open end, P perf or slotted, S screen, T sand point, W walled, X open hole, Z other

TYPE OF SEAL (C67) B bentonite, C clay, G cement grout, N none, Z other

BOTTOM OF SEAL (C68)

METHOD OF DEVELOPMENT (C69) A air-lift pump, B bailed, C compres-sed air, J jetted, N none, P pumped, S surged, Z other

HOURS OF DEVELOPMENT (C70)

SPECIAL TREATMENT (C71) C chem-icals, D dry ice, E explo-sives, F defloc-culent, H hydro-frac-turing, M mech-anical, Z other

CONSTRUCTION HOLE DATA (3 sets shown)

RECORD TYPE (C756) HOLE RECORD SEQUENCE NO. (C724) SEQUENCE NO. OF PARENT RECORD (C59)

DEPTH TO TOP OF INTERVAL (C73) . DEPTH TO BOTTOM OF NTERVAL (C74) . DIAMETER OF INTERVAL (C75) .

RECORD SEQUENCE NO. (C724)

DEPTH TO TOP OF INTERVAL (C73) . DEPTH TO BOTTOM OF NTERVAL (C74) . DIAMETER OF INTERVAL (C75) .

RECORD SEQUENCE NO. (C724)

DEPTH TO TOP OF INTERVAL (C73) . DEPTH TO BOTTOM OF NTERVAL (C74) . DIAMETER OF INTERVAL (C75) .

CONSTRUCTION CASING DATA (4 sets shown)

RECORD TYPE (C758) CSNG RECORD SEQUENCE NO. (C725) SEQUENCE NO. OF PARENT RECORD (C59)

DEPTH TO TOP OF CASING (C77) . DEPTH TO BOTTOM OF CASING (C78) . DIAMETER OF CASING (C79) .

4 CASING MATERIAL (C80) CASING THICKNESS (C81) .

RECORD SEQUENCE NO. (C725) SEQUENCE NO. OF PARENT RECORD (C59)

DEPTH TO TOP OF CASING (C77) . DEPTH TO BOTTOM OF CASING (C78) . DIAMETER OF CASING (C79) .

4 CASING MATERIAL (C80) CASING THICKNESS (C81) .

RECORD SEQUENCE NO. (C725) SEQUENCE NO. OF PARENT RECORD (C59)

DEPTH TO TOP OF CASING (C77) . DEPTH TO BOTTOM OF CASING (C78) . DIAMETER OF CASING (C79) .

4 CAS NG MATERIAL (C80) CAS NG THICKNESS (C81) .

RECORD SEQUENCE NO. (C725) SEQUENCE NO. OF PARENT RECORD (C59)

DEPTH TO TOP OF CASING (C77) . DEPTH TO BOTTOM OF CASING (C78) . DIAMETER OF CASING (C79) .

4 CAS NG MATERIAL (C80) CAS NG THICKNESS (C81) .

FOOTNOTE:

4 CAS NG MATERIAL CODES

A	B	C	D	E	F	G	H	I	J	K	L	M	N	P	Q	R	S	T	U	V	W	X	Y	Z	4	6
abs	brick	concrete	copper	PTFE	Fiber-glass	galv. iron	Fiber-glass plastic	wrought iron	Fiber-glass epoxy	PVC thread-ed	glass	other metal	PVC glued	PVC or plastic	FEP	rock or stone	steel	tile	coated steel	stain-less steel	wood	steel carbon	steel galva-nized	other mat.	stain-less 304	stain-less 316

CONSTRUCTION OPENINGS DATA (3 sets shown)

RECORD TYPE (C760) O P E N — RECORD SEQUENCE NO. (C726) — SEQUENCE NO. OF PARENT RECORD (C59)

DEPTH TO TOP OF INTERVAL (C83) — DEPTH TO BOTTOM OF INTERVAL (C84) — DIAMETER OF INTERVAL (C87)

[5] MATERIAL TYPE (C86) — [6] TYPE OF OPENING (C85) — LENGTH OF OPENING (C89) — WIDTH OF OPENING (C88)

RECORD SEQUENCE NO. (C726)

DEPTH TO TOP OF INTERVAL (C83) — DEPTH TO BOTTOM OF INTERVAL (C84) — DIAMETER OF INTERVAL (C87)

[5] MATERIAL TYPE (C86) — [6] TYPE OF OPENING (C85) — LENGTH OF OPENING (C89) — WIDTH OF OPENING (C88)

RECORD SEQUENCE NO. (C726)

DEPTH TO TOP OF INTERVAL (C83) — DEPTH TO BOTTOM OF INTERVAL (C84) — DIAMETER OF INTERVAL (C87)

[5] MATERIAL TYPE (C86) — [6] TYPE OF OPENING (C85) — LENGTH OF OPENING (C89) — WIDTH OF OPENING (C88)

FOOTNOTES:

[5] TYPE OF MATERIAL CODES FOR OPEN SECTIONS

A	B	C	D	E	F	G	H	I	J	K	L	M	N	P	Q	R	S	T	V	W	X	Y	Z	4	6
ABS	brass or bronze	concrete	ceramic	PTFE	fiber-glass	galv. iron	fiber-glass plastic	wrought iron	fiber-glass epoxy	PVC threaded	glass	other metal	PVC glued	PVC	FEP	stainless steel	steel	tile	brick	membrane	steel carbon	steel galvanized	other	stainless 304	stainless 316

[6] TYPE OF OPENINGS CODES

F	L	M	P	R	S	T	W	X	Z
fractured rock	louvered or shutter-type	mesh screen	perforated, porous or slotted	wire-wound screen	screen (unk.)	sand point screen	walled or shored	open hole	other

CONSTRUCTION MEASURING POINT DATA

RECORD TYPE (C766) M P N T — RECORD SEQUENCE NO. (C728) — BEGINNING DATE (C321) month – day – year — ENDING DATE (C322)

M.P. HEIGHT (C323) — ALTITUDE OF MEASURING POINT (C325) — ALTITUDE METHOD (C326) — ALTITUDE ACCURACY (C327)

ALTITUDE DATUM (C328) — M.P. REMARKS (C324)

RECORD READY FOR WEB (C857)

Y	C	P	L
ready to display	conditional	proprietary	local use only

CONSTRUCTION LIFT DATA

RECORD TYPE (C752) L I F T

RECORD SEQUENCE NO. (C254)

TYPE OF LIFT (C43): A air, B bucket, C centri-fugal, J jet, P piston, R rotary, S submer-sible, T turbine, U un-known, X no lift, Z other

DATE RECORDED (C38) month – day – year

PUMP NTAKE DEPTH (C44)

TYPE OF POWER (C45): D diesel, E electric, G gaso-line, H hand, L LP gas, N natural gas, S solar, W windmill, Z other

HORSE-POWER RATING (C46)

MANUFACTURER (C48)

SERIAL NO. (C49)

POWER COMPANY (C50)

POWER COMPANY ACCOUNT NUMBER (C51)

POWER METER NUMBER (C52)

PUMP RAT NG (C53) (million gallons/units of fuel)

ADDITIONAL LIFT (C255)

PERSON OR COMPANY MAINTAIN NG PUMP (C54)

RATED PUMP CAPACITY (gpm) (C268)

STANDBY POWER (C56) (see TYPE OF POWER)

HORSEPOWER OF STANDBY POWER SOURCE (C57)

MISCELLANEOUS OWNER DATA

RECORD TYPE (C768) O W N R

RECORD SEQUENCE NO. (C718)

DATE OF OWNERSH P (C159)

WU OWNER TYPE (C350): CP Corporation, GV Govern-ment, IN Individual, MI Military, OT Other, TG Tribal, WS Water Supplier

END DATE OF OWNERSHIP (C374)

OWNER'S NAME (C161)

EXAMPLES: JONES, RALPH A.
JONES CONSTRUCTION COMPANY

OWNER'S PHONE NUMBER (C351)

ACCESS TO OWNER'S NAME (C352): 0 Public Access, 1 Coop-erator, 2 USGS Only, 3 District Only, 4 Proprietary

OWNER'S ADDRESS (LINE 1) (C353)

OWNER'S ADDRESS (LINE 2) (C354)

OWNER'S CITY NAME (C355)

STATE (C356)

OWNER'S Z P CODE (C357)

OWNER'S COUNTRY NAME (C358)

ACCESS TO OWNER'S PHONE/ADDRESS (C359): 0 Public Access, 1 Coop-erator, 2 USGS Only, 3 District Only, 4 Proprietary

MISCELLANEOUS VISIT DATA

RECORD TYPE (C774) V I S T

RECORD SEQUENCE NO. (C737)

DATE OF VISIT (C187) month – day – year

NAME OF PERSON (C188)

MISCELLANEOUS OTHER ID DATA (2 sets shown)

RECORD TYPE (C770) O T I D

RECORD SEQUENCE NO. (C736) OTHER ID (C190)

ASSIGNER (C191)

RECORD SEQUENCE NO. (C736) OTHER ID (C190)

ASSIGNER (C191)

MISCELLANEOUS OTHER DATA

RECORD TYPE (C772) O T D T RECORD SEQUENCE NO. (C312)

OTHER DATA TYPE (C181)

OTHER DATA LOCATION (C182) C Cooperator's Office, D District Office R Reporting Agency Z other

DATA FORMAT (C261) F files, M machine readable, P published, Z other

MISCELLANEOUS LOGS DATA (3 sets shown)

RECORD TYPE (C778) L O G S RECORD SEQUENCE NO. (C739) TYPE OF LOG (C199)

BEG NN NG DEPTH (C200) . ENDING DEPTH (C201) . SOURCE OF DATA (C202) A other gov't D driller G geologist L logs M memory O owner R other reported S reporting agency Z other

DATA FORMAT (C225) F files M machine readable P published Z other OTHER DATA LOCATION (C226)

RECORD TYPE (C778) L O G S RECORD SEQUENCE NO. (C739) TYPE OF LOG (C199)

BEG NNING DEPTH (C200) . ENDING DEPTH (C201) . SOURCE OF DATA (C202) A other gov't D driller G geologist L logs M memory O owner R other reported S reporting agency Z other

DATA FORMAT (C225) F files M machine readable P published Z other OTHER DATA LOCATION (C226)

RECORD TYPE (C778) L O G S RECORD SEQUENCE NO. (C739) TYPE OF LOG (C199)

BEGINNING DEPTH (C200) . ENDING DEPTH (C201) . SOURCE OF DATA (C202) A other gov't D driller G geologist L logs M memory O owner R other reported S reporting agency Z other

DATA FORMAT (C225) F files M machine readable P published Z other OTHER DATA LOCATION (C226)

ACOUSTIC LOG:
AS Sonic
AV Acoustic velocity
AW Acoustic waveform
AT Acous ic televiewer

CALIPER LOG:
CP Caliper
CS Caliper, single arm
CT Caliper, three arm
CM Caliper, multi arm
CA Caliper, acous ic

DRILLING LOG:
DT Drilling ime
DR Drillers
DG Geologists
DC Core

ELECTRIC LOG:
EE Electric
ER Single-point resistance
EP Spontaneous potential
EL Long-normal resis ivity
ES Short-normal resis ivity
EF Focused resistivity
ET Lateral resistivity
EN Microresistivity
EC Microresistivity, forused
EO Microresis ivity, lateral
ED Dipmeter

ELECTROMAGNETIC LOG:
MM Magnetic log
MS Magnetic susceptibiity log
MI Electromagne ic induc ion log
MD Electromagne ic dual induction log
MR Radar reflection image log
MV Radar direct-wave velocity log
MA Radar direct-wave amplitude log

FLUID LOG:
FC Fluid conductivity
FR Fluid resis ivity
FT Fluid temperature
FF Fluid differential temperature
FV Fluid velocity
FS Spinner flowmeter
FH Heat-pulse flowmeter
FE Electromagnetic flowmeter
FD Doppler flowmeter
FA Radioactive tracer
FY Dye tracer
FB Brine tracer

NUCLEAR LOG:
NG Gamma
NS Spectral gamma
NA Gamma-gamma
NN Neutron
NT Neutron activitation
NM Neuclear magnetic resonance

OPTICAL LOG:
OV Video
OF Fisheye video
OS Sidewall video
OT Optical televiewer

COMBINATION LOG:
ZF Gamma, fluid resistivity, temperature
ZI Gamma, electromagne ic induction
ZR Long/short normal resistivity
ZT Fluid resistivity, temperature
ZM Electromagnetic flowmeter, fluid resistivity, temperature
ZN Long/short normal resistivity, spontaneous poten ial
ZP Single-point resistance, spontaneous potential
ZE Gamma, long/short normal resistivity, spontaneous potential, single-point resistance, fluid resi ivity, temperature

WELL CONSTRUCTION LOG:
WC Casing collar
WD Borehold deviation

OTHER LOG:
OR Other

MISCELLANEOUS NETWORK DATA (3 types shown)

RECORD TYPE (C780) N E T W RECORD SEQUENCE NO. (C730) TYPE OF NETWORK (C706) Q W water quality BEGINNING YEAR (C115) ENDING YEAR (C116)

TYPE OF ANALYSIS (C120)

A	B	C	D	E	F	G	H	I	J	K	L	M	N	P	Z
physical properties	common ions	trace elements	pesticides	nutrients	sanitary analysis	codes D&B	codes B&E	codes B&C	codes B&F	codes D&E	codes C,D&E	all or most	codes B&C& radioactive	codes B,C&A	other

SOURCE AGENCY (C117) [7]FREQUENCY OF COLLECTION (C118) ANALYZING AGENCY (C307) [8]PRIMARY NETWORK SITE (C257) [8]SECONDARY NETWORK SITE (C708)

RECORD TYPE (C780) N E T W RECORD SEQUENCE NO. (C730) TYPE OF NETWORK (C706) W L water level BEGINNING YEAR (C115) ENDING YEAR (C116)

SOURCE AGENCY (C117) [7] FREQUENCY OF COLLECTION (C118) [8] PRIMARY NETWORK SITE (C257) [8] SECONDARY NETWORK SITE (C708)

RECORD TYPE (C780) N E T W RECORD SEQUENCE NO. (C730) TYPE OF NETWORK (C706) W D pumpage or withdrawals BEGINNING YEAR (C115) ENDING YEAR (C116)

SOURCE AGENCY (C117) [7]FREQUENCY OF COLLECTION (C118) METHOD OF COLLECTION (C133) [8] PRIMARY NETWORK SITE (C257) [8]SECONDARY NETWORK SITE (C708)

C	E	M	U	Z
calculated	estimated	metered	unknown	other

FOOTNOTES:

[7]FREQUENCY OF COLLECTION CODES

A	B	C	D	F	I	M	O	Q	S	W	Z	2	3	4	5	X
annually	bi monthly	continuously	daily	semi-monthly	inter mittent	monthly	one-time only	quarterly	semi-annually	weekly	other	bi-annually	every 3 years	every 4 years	every 5 years	every 10 years

[8]NETWORK SITE CODES

1	2	3	4
national,	district,	project,	co-operator,

MISCELLANEOUS REMARKS DATA (4 types shown)

RECORD TYPE (C788) R M K S RECORD SEQUENCE NO. (C311) DATE OF REMARK (C184) month – day – year

REMARKS (C185)

Subsequent entries may be used to continue the remark. Miscellaneous remarks field is limited to 256 characters.

RECORD TYPE (C788) R M K S RECORD SEQUENCE NO. (C311) DATE OF REMARK (C184) month – day – year

REMARKS (C185)

Subsequent entries may be used to continue the remark. Miscellaneous remarks field is limited to 256 characters.

DISCHARGE DATA

RECORD SEQUENCE NO. (C147)

DATE DISCHARGE MEASURED (C148) month – day – year

TYPE OF DISCHARGE (C703) P pumped, F flow

DISCHARGE (gpm) (C150) .

ACCURACY OF DISCHARGE MEASUREMENT (C310) E excellent (LT 2%), G good (2%-5%), F fair (5%-8%), P poor (GT 8%)

SOURCE OF DATA (C151) A other gov't, D driller, G geologist, L logs, M memory, O owner, R other reported, S reporting agency, Z other

METHOD OF DISCHARGE MEASUREMENT (C152) A acoustic meter, B bailer, C current meter, D Doppler meter, E estimated, F flume, M totaling meter, O orifice, P pitot-tube, R reported, T trajectory, U venturi meter, V volumetric meas, W weir, X unknown, Z other

PRODUCTION WATER LEVEL (C153) .

STATIC WATER LEVEL (C154) .

SOURCE OF DATA (C155) A other gov't, D driller, G geologist, L logs, M memory, O owner, R other reported, S reporting agency, Z other

METHOD OF WATER LEVEL MEASUREMENT (C156) A airline, B recorder, C calibrated airline, E estimated, G pressure gage, H calibrated press. gage, L geophysical logs, M manometer, N non-rec. gage, R reported, S steel tape, T electric tape, U unknown, V calibrated elec. tape, Z other

PUMPING PERIOD (C157) .

SPECIFIC CAPACITY (C272) .

DRAWDOWN (C309) .

GEOHYDROLOGIC DATA

RECORD TYPE (C748) GEOH

RECORD SEQUENCE NO. (C721)

DEPTH TO TOP OF UNIT (C91) .

DEPTH TO BOTTOM OF UNIT (C92) .

UNIT IDENTIFIER (C93)

LITHOLOGY (C96)

CONTRIBUTING UNIT (C304) P principal aquifer, S secondary aquifer, N no contribution, U unknown

LITHOLOGIC MODIFIER (C97)

GEOHYDROLOGIC AQUIFER DATA

RECORD TYPE (C750) AQFR

RECORD SEQUENCE NO. (C742)

SEQUENCE NO. OF PARENT RECORD (C256)

DATE (C95) month – day – year

STATIC WATER LEVEL (C126) .

CONTRIBUTION (C132)

SITE LOCATION SKETCH AND DIRECTIONS

Township ____________ Range ____________

Section # ____________

GWPD 9—Recording minimum and maximum water levels

VERSION: 2010.1

PURPOSE: To determine the minimum and maximum water level in a well between site visits.

Materials and Instruments

1. Plastic spool of nylon fishing leader, 15- or 18-pound test
2. Standard 2 1/2-inch water-level float
3. Transparent 3/8-inch polyethylene tubing
4. Powdered cork
5. Brass tubing, 1/4-inch inside diameter
6. Non-lead shot pellets
7. Hammer, nails, and screw-eye hooks
8. Hacksaw
9. Graduated steel tape
10. Permanent, water-resistant marker
11. Field notebook
12. Pencil or pen, blue or black ink. Strikethrough, date, and initial errors; no erasures
13. Safety equipment: gloves, safety glasses, first-aid kit

Data Accuracy and Limitations

1. Devices were tested in a well having a continuous recorder and found to measure water levels to an accuracy of 0.1 foot.
2. Use should be limited to wells with water-level depths of 50 feet or less.
3. The well diameter is limited to 3 inches or larger with a standard 2 1/2-inch water-level float. In smaller diameter wells, a weighted dowel could be used in place of the standard float.

Advantages

1. Three water-level measurements can be obtained for each visit to the site regardless of the length of time between visits.
2. Devices are inexpensive and easy to install.
3. Devices can last indefinitely.

Disadvantages

1. If kinks occur in the polyethylene tubing, they may prevent the movement of the powdered cork and could cause anomalous readings.
2. If these devices are used in wells with water levels deeper than 50 feet, the nylon leader may stretch and give anomalous readings.
3. Dates of the minimum and maximum water levels cannot be determined.

Assumptions

1. No continuous recorder is available or necessary.
2. Dates of the maximum and minimum water levels are not critical.
3. The well has a shelter that contains a wooden base or subfloor.

Instructions

1. Construct the device for measuring maximum water levels (fig. 1, items 1–4).

   a. The maximum water-level device consists of a length of transparent 3/8-inch polyethylene tubing, two lengths of 1/4-inch inside diameter brass tubing, non-lead shot, powdered cork, and a nail.

   b. Crimp one end of an 8- to 12-inch length of brass tubing, slot the brass tubing with a hacksaw over the lower 3/4 of its length, fill the brass tubing with non-lead shot, and attach it to the lower end of the polyethylene tubing. Be sure to place enough non-lead shot in the polyethylene tubing so that the tubing hangs taut in the well and contains no kinks. The length of polyethylene tubing selected must be long enough to keep the lower 12 inches of the brass tubing submerged below the water surface at all times.

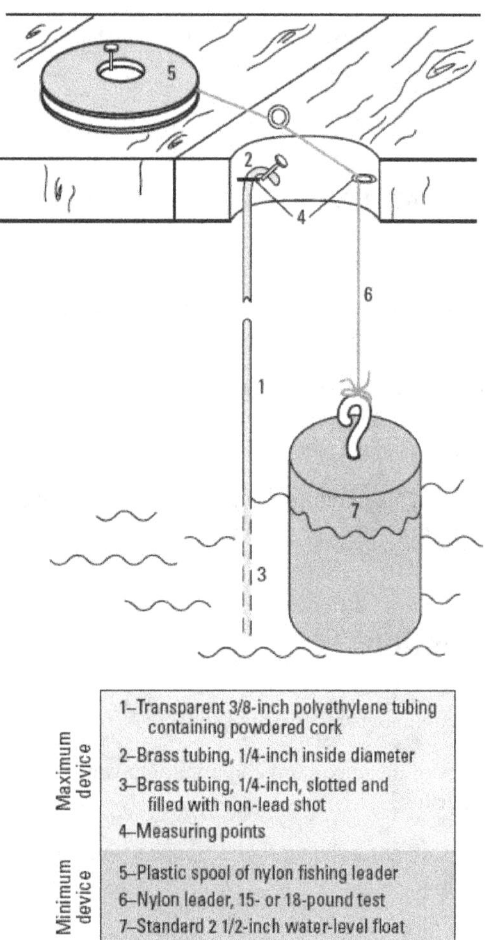

Figure 1. Devices for measuring maximum and minimum water levels in wells (modified from Kelly, 1968).

   c. Put several pinches of powdered cork in the polyethylene tubing.

   d. Bend a short length of brass tubing to form an elbow and insert the brass elbow into the upper end of the polyethylene tubing.

   e. Insert a nail in the wood base or subfloor of the well shelter to use as a measuring point. Mark the measuring point on the tubing with the permanent marker.

   f. Suspend the maximum water-level device in the well by hanging the brass elbow over the measuring point nail.

2. Determine the maximum water level for the well. The powdered cork adheres to the walls of the polyethylene tubing as the water level in the well rises, thereby marking the maximum water level. The maximum water-level device is a modification of a crest-stage gage.

   a. Gently withdraw the tubing assembly from the well.

   b. Measure the distance between the measuring point and the top of the powdered cork with a graduated steel tape.

   c. Record the maximum water level in the field notebook.

   d. Shake the powdered cork to the bottom of the device and re-install the maximum water-level device.

3. Construct the device for measuring minimum water levels (fig. 1, items 5–7).

   a. The minimum water-level device consists of nylon fishing leader wound on a disc-shaped spool, a standard 2 1/2-inch water-level float, a nail, and two screw-eye hooks.

   b. Attach the disc-shaped spool to the wooden base or shelter subfloor with a nail.

   c. Attach the two screw-eye hooks to the subfloor as shown in figure 1. The lower eye hook is used as a measuring point.

   d. Thread the nylon fishing leader from the disc-shaped spool through the screw-eye hooks and secure the nylon leader to the top of the float.

   e. Mark the waterline on the float with a permanent, water resistant marker before installing the float in the well.

4. Determine the minimum water level for the well. The water-level float pulls the nylon fishing leader from the spool as the water level declines and the nylon leader becomes slack. Spool friction prevents the nylon leader from rewinding.

    a. Place the nail of the index finger on the nylon leader at the eye hook measuring point to mark the leader.

    b. Hold your index finger on the leader mark and gently withdraw the nylon leader from the well.

    c. Measure the amount of nylon leader between the measuring point and the float plus the distance from the float-leader connection to the float waterline with a graduated steel tape.

    d. Record the minimum water level in the field notebook.

    e. Rewind the spool and re-install the minimum water-level device.

Data Recording

Record minimum and maximum water levels in the field notebook.

References

Kelly, T.E., 1968, Minimum and maximum water-level recording devices, *in* Chase, E.B., and Payne, F.N., comps., Selected techniques in water resources investigations, 1966–67: U.S. Geological Survey Water-Supply Paper 1892, p. 83–86.

GWPD 10—Estimating discharge from a pumped well by use of a circular orifice weir

VERSION: 2010.1

PURPOSE: To estimate the discharge from a pumped well from a non-vertical standard pipe by using a circular orifice weir.

Materials and Instruments

1. Steel orifice plate
2. Hand level
3. Piezometer tube, 1/8-inch or 1/4-inch diameter
4. Glass tube, 1/8-inch or 1/4-inch diameter
5. Accurate yardstick, or other suitable ridged scale
6. Graduated tape
7. Pencil or pen, blue or black ink. Strikethrough, date, and initial errors; no erasures
8. Field notebook
9. Groundwater Site Inventory (GWSI) System Groundwater Site Schedule, Form 9-1904-A

Data Accuracy and Limitations

1. The circular orifice weir method is accurate to within 2 percent.
2. The hole in the steel plate of the orifice weir must be accurately cut, be centered, be circular, and have a beveled edge. The steel plate restricts the flow through the orifice and creates a pressure head in the discharge pipe.
3. For the orifice weir to function properly, the gate valve that controls the rate of discharge must be placed at least 10 pipe diameters from the piezometer tube connection to keep pipe turbulence to a minimum.
4. The piezometer tube must be completely free of any obstruction and free of air bubbles when a reading of the pressure head is made. The head in the line is correlated with discharge by use of tables calibrated for the particular ratio between the orifice and the discharge pipe diameters (table 1).
5. The discharge pipe must be level, and the water flow from the end of the discharge pipe must fall freely.

Advantages

1. This method provides an accurate means of determining the discharge rate from turbine or centrifugal pumps.
2. No special training is needed to use this method.

Disadvantages

1. This method cannot be used to measure the pulsating flow from a piston pump.
2. Well flow must be constant.

Assumptions

1. An appropriately sized orifice plate is available and was built accurately.
2. The diameter of the orifice plate is less than eight-tenths of the inside diameter of the pipe that serves as the channel of approach.
3. The last 6 feet of the discharge line is level and contains a fitting that is screwed into a 1/8-inch or 1/4-inch tapped hole centered on the discharge line, exactly 24 inches from the orifice plate.

Instructions

1. Figure 1 shows the essential details for setting up a circular orifice weir for measuring the discharge rate of a well that is being pumped with a turbine or centrifugal pump.

2. Select an appropriately sized circular orifice weir and attach it to the end of the discharge pipe. Table 1 lists 3- to 10-inch circular orifice weirs that can be used with discharge pipes ranging from 4- to 12-inches in diameter.

3. Place a short piece of glass tubing into the upper end of the piezometer tube. Attach the lower end of the piezometer tube to the fitting on the discharge line that is located 24 inches from the orifice plate (fig. 1). Tape the piezometer tube to the scale making sure that the zero mark on the scale lines up with the center of the piezometer fitting in the discharge pipe.

4. The water level in the piezometer tube represents the pressure in the approach pipe when water is being pumped through the orifice. The water level can be observed in the glass tube.

5. To read the pressure head in the glass tube, hold the piezometer tube in an upright position perpendicular to the discharge pipe. Read the water level using the attached scale.

6. Determine the well discharge from table 1. For example, if the pressure head is 25.5 inches, the orifice plate is 5 inches in diameter and the discharge pipe is 8 inches in diameter; follow the 25.5-inch line from the left scale until it intersects with the 5-inch orifice and 8-inch pipe column. The well discharge rate obtained from table 1 is 500 gallons per minute.

7. Between water-level readings, check for air bubbles in the piezometer tube. If air bubbles are present, they can be eliminated from the piezometer tube by dropping the tube between readings so that water flows from it.

8. Record estimated discharge in the field notebook and in the discharge data section of the GWSI Groundwater Site Schedule (fig. 2, Form 9-1904-A).

Data Recording

Data are recorded in a field notebook. Discharge data should also be recorded in the discharge data section of the GWSI Groundwater Site Schedule (Form 9-1904-A).

References

Driscoll, F.G., 1986, Groundwater and wells (2d ed.): St. Paul, Minnesota, Johnson Filtration Systems, Inc., 1089 p.

Hoopes, B.C., ed., 2004, User's manual for the National Water Information System of the U.S. Geological Survey, Ground-Water Site-Inventory System (version 4.4): U.S. Geological Survey Open-File Report 2005–1251, 274 p.

Layne & Bowler, Inc., 1958, Measurement of water flow through pipe orifice with free discharge: Memphis, TN, Layne & Bowler, Inc., Bulletin 501, p. 22–25.

U.S. Geological Survey, Office of Water Data Coordination, 1977, National handbook of recommended methods for water-data acquisition: Office of Water Data Coordination, Geological Survey, U.S. Department of the Interior, chap. 2, p. 2-17.

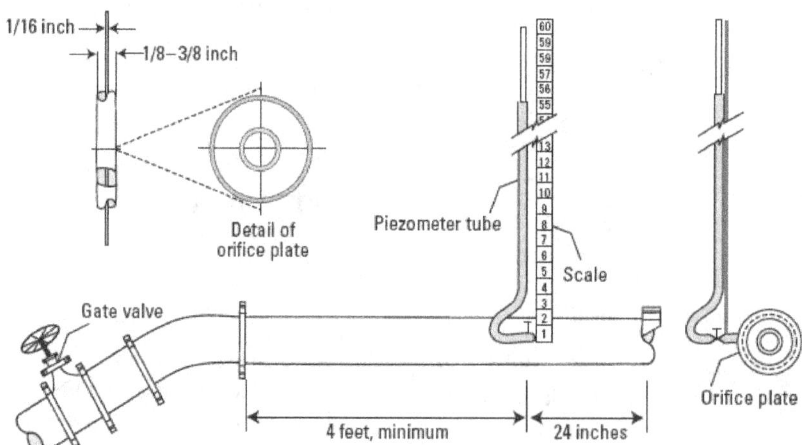

Figure 1. Essential details of the circular orifice weir commonly used for measuring well discharge when pumping by means of a turbine pump. Discharge pipe must be level (Driscoll, 1986).

Table 1. Orifice table for measurement of water through pipe orifices with free discharge. Values are in gallons per minute to the nearest whole number. (Compiled by the Engineering Department of Layne and Bowier, Inc., from original calibrations by Purdue University)

[—; no data]

Head, in inches	3-inch orifice		4-inch orifice		5-inch orifice		6-inch orifice		7-inch orifice	8-inch orifice	9-inch orifice	10-inch orifice
	4-inch pipe	6-inch pipe	6-inch pipe	8-inch pipe	6-inch pipe	8-inch pipe	8-inch pipe	10-inch pipe	10-inch pipe	10-inch pipe	12-inch pipe	12-inch pipe
5	100	76	145	140	280	220	380	320	—	—	825	1,100
5.5	104	79	153	145	293	230	394	333	—	—	860	1,150
6	108	82	160	150	305	240	408	345	—	—	895	1,200
6.5	111	85	167	155	316	250	421	358	—	—	930	1,250
7	115	88	172	160	328	260	433	370	—	—	965	1,300
7.5	119	91	179	165	339	270	446	383	—	—	1,000	1,350
8	122	94	185	170	350	280	458	395	600	935	1,032	1,400
8.5	125	96	190	175	361	289	471	408	617	963	1,065	1,440
9	128	99	195	180	372	298	483	420	633	992	1,093	1,480
9.5	130	102	200	185	383	307	495	433	650	1,016	1,120	1,520
10	133	104	205	190	393	316	508	445	666	1,040	1,148	1,560
10.5	137	107	210	195	402	324	521	458	682	1,060	1,172	1,600
11	140	109	215	200	412	330	533	470	698	1,080	1,200	1,635
11.5	143	111	220	204	421	338	545	480	713	1,100	1,225	1,670
12	146	114	225	208	430	346	556	490	728	1,120	1,250	1,705
12.5	149	116	230	212	439	354	567	500	743	1,139	1,277	1,740
13	151	118	234	216	448	362	578	510	757	1,158	1,303	1,775
13.5	154	121	239	219	457	369	589	520	771	1,176	1,328	1,810
14	157	123	243	224	465	376	599	530	785	1,194	1,352	1,845
14.5	159	126	247	227	473	383	609	540	799	1,212	1,376	1,875
15	162	128	250	231	480	390	618	550	812	1,230	1,400	1,905
15.5	164	130	254	234	488	396	627	559	825	1,248	1,421	1,940
16	167	132	257	238	495	402	636	568	838	1,266	1,441	1,970
16.5	170	134	261	241	503	408	645	577	851	1,284	1,460	2,000
17	172	136	264	245	510	414	654	586	863	1,302	1,480	2,030
17.5	175	138	268	249	517	420	663	595	875	1,319	1,500	2,060
18	178	140	271	252	524	426	672	604	887	1,336	1,520	2,089
18.5	180	142	275	256	530	432	681	612	899	1,353	1,540	2,118
19	183	144	278	259	536	438	690	620	910	1,370	1,560	2,146
19.5	185	146	282	263	542	444	699	628	922	1,387	1,580	2,175
20	187	148	285	266	548	449	708	636	933	1,404	1,600	2,204
20.5	190	150	289	270	554	455	717	643	945	1,421	1,620	2,232
21	192	152	292	273	560	460	726	650	956	1,438	1,640	2,260
21.5	195	154	295	275	566	465	735	657	968	1,455	1,659	2,288
22	197	156	299	279	572	470	744	664	979	1,471	1,677	2,316
22.5	199	158	302	282	578	475	752	671	990	1,486	1,695	2,343
23	201	160	305	285	584	479	760	678	1,001	1,500	1,714	2,360
23.5	203	162	307	288	590	484	768	685	1,012	1,515	1,732	2,382
24	205	164	310	291	596	488	776	692	1,022	1,529	1,750	2,409
24.5	207	165	314	294	602	492	784	699	1,033	1,543	1,767	2,435
25	210	167	317	297	608	496	791	706	1,043	1,557	1,783	2,461
25.5	212	169	320	300	614	500	798	713	1,059	1,571	1,799	2,487
26	214	171	323	303	620	504	805	720	1,064	1,585	1,815	2,513
26.5	216	173	326	305	626	508	812	727	1,074	1,599	1,830	2,539
27	219	174	329	308	632	512	818	734	1,084	1,613	1,845	2,565

Table 1. Orifice table for measurement of water through pipe orifices with free discharge. Values are in gallons per minute to the nearest whole number. (Compiled by the Engineering Department of Layne and Bowier, Inc., from original calibrations by Purdue University)—Continued

[—; no data]

Head, in inches	3-inch orifice		4-inch orifice		5-inch orifice		6-inch orifice		7-inch orifice	8-inch orifice	9-inch orifice	10-inch orifice
	4-inch pipe	6-inch pipe	6-inch pipe	8-inch pipe	6-inch pipe	8-inch pipe	8-inch pipe	10-inch pipe	10-inch pipe	10-inch pipe	12-inch pipe	12-inch pipe
27.5	221	176	332	311	638	516	825	741	1,094	1,627	1,860	2,590
28	222	177	335	314	644	520	831	747	1,104	1,641	1,875	2,610
28.5	224	179	337	317	650	524	838	754	1,114	1,655	1,890	2,630
29	226	180	340	320	656	528	844	760	1,124	1,669	1,905	2,650
29.5	228	182	343	323	662	532	851	767	1,134	1,683	1,920	2,670
30	230	183	346	325	668	536	857	773	1,143	1,697	1,935	2,690
30.5	232	185	348	328	674	540	863	780	1,153	1,711	1,950	2,713
31	235	186	351	330	680	544	869	786	1,162	1,725	1,965	2,736
31.5	236	188	354	333	686	548	876	793	1,172	1,739	1,980	2,759
32	239	189	357	335	692	552	882	799	1,181	1,753	2,005	2,782
32.5	240	191	360	338	697	556	889	806	1,191	1,767	2,020	2,805
33	242	192	363	340	703	560	895	812	1,200	1,791	2,040	2,828
33.5	244	194	366	342	709	564	901	818	1,209	1,795	2,050	2,850
34	246	195	369	345	715	568	907	824	1,218	1,809	2,060	2,873
34.5	248	196	372	247	720	572	913	830	1,227	1,823	2,075	2,896
35	250	197	375	349	726	576	919	836	1,235	1,837	2,090	2,919
35.5	252	198	377	351	732	580	925	842	1,243	1,851	2,100	2,941
36	254	200	380	354	737	584	931	847	1,251	1,865	2,112	2,964
36.5	256	201	383	356	743	588	937	852	1,259	1,879	2,124	2,980
37	257	203	385	358	748	592	943	857	1,266	1,893	2,136	3,002
37.5	259	204	388	360	754	596	949	862	1,274	—	2,148	3,024
38	260	205	390	363	759	600	955	867	1,281	—	2,160	3,046
38.5	262	206	393	365	765	604	961	872	1,289	—	2,173	3,068
39	263	208	396	367	770	608	967	877	1,295	—	2,185	3,088
39.5	265	209	398	369	776	612	974	882	1,304	—	2,197	3,110
40	266	210	401	371	781	616	979	887	1,311	—	2,210	3,130
40.5	267	211	403	373	786	620	985	891	1,319	—	2,225	3,146
41	269	212	406	375	790	624	990	896	1,326	—	2,233	3,160
41.5	271	213	408	378	795	628	996	901	1,334	—	2,245	3,179
42	272	214	411	380	800	631	1001	906	1,341	—	2,257	3,199
42.5	274	216	413	382	805	635	1007	910	1,349	—	2,273	3,219
43	275	217	415	384	810	638	1012	915	1,356	—	2,285	3,230
43.5	277	218	418	386	815	642	1018	920	1,364	—	2,397	3,250
44	278	219	420	388	820	645	1023	925	1,371	—	2,309	3,263
44.5	280	220	422	390	824	649	1029	929	1,379	—	2,326	3,280
45	281	222	425	392	828	652	1034	934	1,387	—	2,338	3,298
45.5	283	223	427	394	832	656	1040	939	1,394	—	2,350	3,316
46	284	224	429	396	837	659	1045	944	1,401	—	2,363	3,334
46.5	285	225	432	399	842	663	1051	948	1,409	—	2,375	3,351
47	287	227	434	401	847	666	1056	953	1,416	—	2,387	3,368
47.5	289	228	437	403	851	669	1062	958	1,424	—	2,399	3,389
48	290	229	440	405	855	672	1067	963	1,431	—	2,411	3,405
48.5	292	230	442	407	859	676	1073	967	1,439	—	2,423	3,426
49	293	231	444	409	863	679	1078	972	1,446	—	2,434	3,443
49.5	294	232	446	411	868	683	1084	977	1,454	—	2,444	3,460

Table 1. Orifice table for measurement of water through pipe orifices with free discharge. Values are in gallons per minute to the nearest whole number. (Compiled by the Engineering Department of Layne and Bowier, Inc., from original calibrations by Purdue University)—Continued

[—; no data]

Head, in inches	3-inch orifice		4-inch orifice		5-inch orifice		6-inch orifice		7-inch orifice	8-inch orifice	9-inch orifice	10-inch orifice
	4-inch pipe	6-inch pipe	6-inch pipe	8-inch pipe	6-inch pipe	8-inch pipe	8-inch pipe	10-inch pipe	10-inch pipe	10-inch pipe	12-inch pipe	12-inch pipe
50	296	234	448	413	872	686	1089	982	1,461	—	2,454	3,477
50.5	298	235	450	415	876	690	1095	986	1,469	—	2,464	3,494
51	300	236	453	417	880	693	1100	991	1,476	—	2,474	3,511
51.5	301	237	455	419	884	697	1105	996	1,484	—	2,486	3,527
52	302	238	457	421	888	700	1110	1000	1,491	—	2,498	3,544
52.5	303	239	459	423	892	704	1115	1005	1,499	—	2,510	3,560
53	304	240	461	425	896	707	1,120	1,009	1,506	—	2,522	3,575
53.5	305	241	463	427	900	711	1,125	1,014	1,513	—	2,534	3,591
54	307	243	465	429	904	714	1,130	1,018	1,520	—	2,545	3,602
54.5	309	244	467	431	908	718	1,135	1,023	1,527	—	2,555	3,618
55	310	246	469	433	912	721	1,140	1,027	1,534	—	2,565	3,634
55.5	311	247	471	435	915	725	1,145	1,032	1,541	—	2,575	3,650
56	313	248	472	437	919	727	1,150	1,036	1,548	—	2,586	3,667
56.5	314	249	474	439	923	730	1,155	1,040	1,554	—	2,597	3,684
57	315	250	476	441	927	733	1,160	1,044	1,560	—	2,608	3,702
57.5	316	251	478	443	930	736	1,165	1,046	1,567	—	2,619	3,719
58	317	252	480	445	934	739	1,170	1,052	1,574	—	2,630	3,736
58.5	319	253	482	447	938	742	1,175	1,056	1,580	—	2,641	3,752
59	320	254	485	449	942	745	1,180	1,060	1,586	—	2,653	3,768
59.5	321	256	487	451	945	748	1,185	1,064	1,592	—	2,665	3,784
60	323	257	489	453	948	751	1,190	1,068	1,598	—	2,676	3,800
60.5	324	258	491	455	951	754	1,195	1,072	—	—	—	—
61	325	259	492	457	955	757	1,200	1,076	—	—	—	—
61.5	326	261	494	459	958	760	1,205	1,080	—	—	—	—
62	328	262	496	461	961	763	1,209	1,084	—	—	—	—
62.5	329	263	498	463	964	766	1,214	1,088	—	—	—	—
63	330	264	500	465	968	769	1,218	1,092	—	—	—	—
63.5	331	265	502	467	971	772	1,223	1,096	—	—	—	—
64	333	266	504	469	974	775	1,227	1,099	—	—	—	—
64.5	334	267	507	471	977	778	1,232	1,103	—	—	—	—
65	335	268	509	472	981	781	1,236	1,106	—	—	—	—
65.5	336	269	511	474	984	784	1,241	1,110	—	—	—	—
66	338	271	513	475	988	787	1,245	1,113	—	—	—	—
66.5	339	272	515	477	991	790	1,250	1,117	—	—	—	—
67	340	273	517	479	995	793	1,254	1,120	—	—	—	—
67.5	341	274	518	481	998	796	1,259	1,124	—	—	—	—
68	343	275	520	483	1,002	799	1,263	1,127	—	—	—	—
68.5	344	276	521	485	1,005	802	1,268	1,131	—	—	—	—
69	346	277	523	487	1,009	805	1,272	1,134	—	—	—	—
69.5	347	278	524	489	1,012	808	1,276	1,137	—	—	—	—
70	349	280	525	491	1,016	811	1,280	1,140	—	—	—	—

FORM NO. 9-1904-A
Revised Sept 2009, NWIS 4.9

File Code ____________
Date ____________

Coded by ____________
Checked by ____________
Entered by ____________

U.S DEPT. OF THE INTERIOR
GEOLOGICAL SURVEY

GROUNDWATER SITE SCHEDULE
General Site Data

AGENCY CODE (C4) USGS | SITE ID (C1) | PROJECT (C5)

STATION NAME (C12/900)

[1] SITE TYPE (C802) Primary - Secondary | DISTRICT (C6) | COUNTRY (C41) | STATE (C7)

COUNTY or TOWN (C8) ____________ County code

LATITUDE (C9) . | LONGITUDE (C10) . | LAT/LONG ACCURACY (C11): H (Hndrth sec.), 1 (tenth sec.), 5 (half sec.), S (sec.), R (3 sec.), F (5 sec.), T (10 sec.), M (min.), U (Un-known)

LAT/LONG METHOD (C35): C (land net), D (DGPS), G (GPS), L (LORAN), M (map), N (inter-polated digital map), R (reported), S (survey), U (un-known) | LAT/LONG DATUM (C36): NAD27 (North American Datum of 1927), NAD83 (North American Datum of 1983) | ALTITUDE (C16) .

ALTITUDE ACCURACY (C18) | ALTITUDE METHOD (C17): A (altimeter), D (DGPS), G (GPS), I (IfSAR), J (LIDAR), L (Level), M (map), N (DEM), R (re-ported), U (un-known) | ALTITUDE DATUM (C22): NGVD29 (National Geodetic Vertical Datum of 1929), NAVD88 (North American Vertical Datum of 1988)

LAND NET (C13): ¼ ¼ ¼ S section T township range merid

TOPOGRAPHIC SETTING (C19): A (alluvial fan), B (playa), C (stream channel), D (depres-sion), E (dunes), F (flat), G (flood-plain), H (hill-top), K (sink-hole), L (lake or swamp), M (mangrove swamp), O (off-shore), P (pedi-ment), S (hill-side), T (ter-race), U (undu-lating), V (valley flat), W (upland draw)

HYDROLOGIC UNIT CODE (C20) | DRAINAGE BASIN CODE (C801) | STANDARD TIME ZONE (C813) | DAYLIGHT SAVINGS TIME FLAG (C814) Y OR **N**

MAP NAME (C14) | MAP SCALE (C15)

AGENCY USE (C803): A (active nolna), D (discon-tinued), I (inactive site), L (active written), M (active oral), O (inventory site), R (remediated) | [2] NATIONAL WATER-USE (C39)

DATA TYPE (C804)
Place an 'A' (active), an 'I' (inactive), or an 'O' (inventory) in the appropriate box: WL cont, WL int, QW cont, QW int, PR cont, PR int, EV cont, EV int, wind vel., tide cont, tide int, sed. con, sed. ps, peak flow, low flow, state water use

INSTRUMENTS (C805) (Place a "Y" in the appropriate box): digital rec-order, graphic rec-order, tele-metry land line, tele-metry radio, tele-metry satellite, AHDAS, crest-stage gage, tide gage, deflec-tion meter, bubble gage, stilling well, CR type recorder, weigh-ing rain gage, tipping bucket rain gage, acoustic velocity meter, electro-magnetic flowmeter, pressure transducer

DATE INVENTORIED (C711) month – day – year | RECORD READY FOR WEB (C32): Y (ready to display), C (condi-tional), P (proprie-tary), L (local use only)

REMARKS (C806)

FOOTNOTES

1SITE TYPE (C802)

GL	Glacier	OC	Ocean	GW	Well	SB	Subsurface
WE	Wetland	OC-CO	Coastal	GW-CR	Collector or Ranney type well	SB-CV	Cave
AT	Atmosphere	LK	Lake, Reservoir, Impoundment	GW-EX	Extensometer well	SB-GWD	Groundwater drain
ES	Estuary			GW-HZ	Hyporheic -zone well	SB-TSM	Tunnel, shaft, or mine
LA	Land	SP	Spring	GW-IW	Interconnected wells	SB-UZ	Unsaturated zone
LA-EX	Excavation	ST	Stream	GW-TH	Test hole not completed as a well		
LA-OU	Outcrop	ST-CA	Canal	GW-MW	Multiple wells		
LA-SNK	Sinkhole	ST-DCH	Ditch				
LA-SH	Soil hole	ST-TS	Tidal stream				
LA-SR	Shore	FA-WIW	Waste-Injection well				

[2] WS (water supply), DO (domestic), CO (commer-cial), IN (industrial), IR (irrigation), MI (mining), LV (livestock), PH (power hydro-electric), ST (waste water treatment), RM (remedia-tion), TE (thermo-electric power), AQ (aqua-culture)

C22 Other (see manual for codes)
C36 Other (see manual for codes)
C39 is mandatory for all sites having data in SWUDS.

Figure 2. Groundwater Site Schedule, Form 9-1904-A.

GENERAL SITE DATA

DATA RELIAB LITY (C3): C field checked | L poor location | M minimal data | U un-checked

DATE OF FIRST CONSTRUCTION (C21): month – day – year

USE OF SITE (C23): A anode | C standby emer. supply | D drain | E geo-thermal | G seismic | H heat reservoir | M mine | O obser-vation | P oil or gas | R recharge | S repres-surize | T test | U unused | V with-drawal/ return | W with-drawal | X waste | Z des-troyed

SECOND-ARY USE OF SITE (C301) (See use of site)

TERTIARY USE OF SITE (C302) (See use of site)

USE OF WATER (C24): A air cond. | B bottling | C comm-ercial | D de-water | E power | F fire | H domes-tic | I irri-gation | J indus-trial (cooling) | K mining | M medi-cinal | N indus-trial | P public supply | Q aqua-culture | R recrea-tions | S stock | T insti-tutional | U unused | Y desalin-ation | Z other

SECOND-ARY USE OF WATER (C25) (see use of water)

TERTIARY USE OF WATER (C26) (see use of water)

AQUIFER TYPE (C713): U unconfined single | N unconfined multiple | C confined single | M confined multiple | X mixed

PRIMARY AQU FER (C714)

NATIONAL AQUIFER (C715)

HOLE DEPTH (C27) .

WELL DEPTH (C28) .

SOURCE OF DEPTH DATA (C29): A other gov't | D driller | G geol-ogist | L logs | M memory | O owner | R other reported | S reporting agency | Z other

WATER-LEVEL DATA

DATE WATER-LEVEL MEASURED (C235): month – day – year

TIME (C709)

WATER-LEVEL TYPE CODE (C243): L land surface | M meas. pt. | S vertical datum

WATER LEVEL (C237/241/242) .

MP SEQUENCE NO. (C248) (Mandatory if WL type=M)

WATER-LEVEL DATUM (C245) (Mandatory if WL type=S): NGVD29 National Geodetic Vertical Datum Of 1929 | NAVD88 North American Vertical Datum Of 1988 | Other (See manual for codes)

SITE STATUS FOR WATER LEVEL (C238): A atmos. pressure | B tide stage | C ice | D dry | E recently flowing | F flowing | G nearby flowing | H nearby recently flowing | I injector site | J injector site monitor | M plugged | N measure-ment discontinued | O obstruc-tion | P pumping | R recently pumped | S nearby pumping | T nearby recently pumped | V foreign sub-stance | W well des-troyed | X affected by surface water | Z other

METHOD OF WATER-LEVEL MEASUREMENT(C239): A airline | B analog | C calibrated airline | D differ-ential GPS | E esti-mated | F trans-ducer | G pressure gage | H calibrated press. gage | L geophysi-cal logs | M mano-meter | N non-rec. gage | O observed | P acoustic pulse | R reported | S steel tape | T electric tape | V calibrated elec. tape | Z other

WATER-LEVEL ACCURACY (C276): 0 foot | 1 tenth | 2 hun-dredth | 9 not to nearest foot

SOURCE OF WATER-LEVEL DATA (C244): A other gov't | D driller's log | G geol-ogist | L geophysi-cal logs | M memory | O owner | R other reported | S reporting agency | Z other

PERSON MAK NG MEASUREMENT (C246) (WATER LEVEL PARTY)

MEASUR NG AGENCY (C247) (SOURCE)

EQUIP ID (C249) (20 char)

REMARKS (C267) (256 char)

RECORD READY FOR WEB (C858): Y ready to display | C condi-tional | P proprie-tary | L local use only

CONSTRUCTION DATA

RECORD TYPE (C754): CONS

RECORD SEQUENCE NO. (C723)

DATE OF COMPLETED CONSTRUCTION (C60): month – day – year

NAME OF CONTRACTOR (C63)

SOURCE OF DATA (C64): A other gov't | D driller | G geol-ogist | L logs | M memory | O owner | R other reported | S reporting agency | Z other

METHOD OF CONSTRUCTION (C65): A air-rotary | B bored or augered | C cable tool | D dug | H hydraulic rotary | J jetted | P air per-cussion | R reverse rotary | S sonic | T trenching | V driven | W drive wash | Z other

TYPE OF F NISH (C66): C porous concrete | F gravel w/perf. | G gravel screen | H horiz. gallery | O open end | P perf or slotted | S screen | T sand point | W walled | X open hole | Z other

TYPE OF SEAL (C67): B bentonite | C clay | G cement grout | N none | Z other

BOTTOM OF SEAL (C68)

METHOD OF DEVELOPMENT (C69): A air-lift pump | B bailed | C compres-sed air | J jetted | N none | P pumped | S surged | Z other

HOURS OF DEVELOPMENT (C70)

SPECIAL TREATMENT (C71): C chem-icals | D dry ice | E explo-sives | F defloc-culent | H hydro-frac-turing | M mech-anical | Z other

CONSTRUCTION HOLE DATA (3 sets shown)

RECORD TYPE (C756) HOLE

RECORD SEQUENCE NO. (C724)

SEQUENCE NO. OF PARENT RECORD (C59)

DEPTH TO TOP OF INTERVAL (C73) .

DEPTH TO BOTTOM OF NTERVAL (C74) .

DIAMETER OF INTERVAL (C75) .

RECORD SEQUENCE NO. (C724)

DEPTH TO TOP OF INTERVAL (C73) .

DEPTH TO BOTTOM OF NTERVAL (C74) .

DIAMETER OF INTERVAL (C75) .

RECORD SEQUENCE NO. (C724)

DEPTH TO TOP OF INTERVAL (C73) .

DEPTH TO BOTTOM OF NTERVAL (C74) .

DIAMETER OF INTERVAL (C75) .

CONSTRUCTION CASING DATA (4 sets shown)

RECORD TYPE (C758) CSNG

RECORD SEQUENCE NO. (C725)

SEQUENCE NO. OF PARENT RECORD (C59)

DEPTH TO TOP OF CASING (C77) .

DEPTH TO BOTTOM OF CASING (C78) .

DIAMETER OF CASING (C79) .

4 CASING MATERIAL (C80)

CASING THICKNESS (C81) .

RECORD SEQUENCE NO. (C725)

SEQUENCE NO. OF PARENT RECORD (C59)

DEPTH TO TOP OF CASING (C77) .

DEPTH TO BOTTOM OF CASING (C78) .

DIAMETER OF CASING (C79) .

4 CASING MATERIAL (C80)

CASING THICKNESS (C81) .

RECORD SEQUENCE NO. (C725)

SEQUENCE NO. OF PARENT RECORD (C59)

DEPTH TO TOP OF CASING (C77) .

DEPTH TO BOTTOM OF CASING (C78) .

DIAMETER OF CASING (C79) .

4 CAS NG MATERIAL (C80)

CAS NG THICKNESS (C81) .

RECORD SEQUENCE NO. (C725)

SEQUENCE NO. OF PARENT RECORD (C59)

DEPTH TO TOP OF CASING (C77) .

DEPTH TO BOTTOM OF CASING (C78) .

DIAMETER OF CASING (C79) .

4 CAS NG MATERIAL (C80)

CAS NG THICKNESS (C81) .

FOOTNOTE:

[4] CAS NG MATERIAL CODES

A	B	C	D	E	F	G	H	I	J	K	L	M	N	P	Q	R	S	T	U	V	W	X	Y	Z	4	6
abs	brick	concrete	copper	PTFE	Fiber-glass	galv. iron	Fiber-glass plastic	wrought iron	Fiber-glass epoxy	PVC thread-ed	glass	other metal	PVC glued	PVC or plastic	FEP	rock or stone	steel	tile	coated steel	stain-less steel	wood	steel carbon	steel galva-nized	other mat.	stain-less 304	stain-less 316

CONSTRUCTION OPENINGS DATA (3 sets shown)

RECORD TYPE (C760) OPEN RECORD SEQUENCE NO. (C726) SEQUENCE NO. OF PARENT RECORD (C59)

DEPTH TO TOP OF INTERVAL (C83) . DEPTH TO BOTTOM OF INTERVAL (C84) . DIAMETER OF INTERVAL (C87) .

[5] MATERIAL TYPE (C86) [6] TYPE OF OPENING (C85) LENGTH OF OPENING (C89) . WIDTH OF OPENING (C88) .

RECORD SEQUENCE NO. (C726)

DEPTH TO TOP OF INTERVAL (C83) . DEPTH TO BOTTOM OF INTERVAL (C84) . DIAMETER OF INTERVAL (C87) .

[5] MATERIAL TYPE (C86) [6] TYPE OF OPENING (C85) LENGTH OF OPENING (C89) . WIDTH OF OPENING (C88) .

RECORD SEQUENCE NO. (C726)

DEPTH TO TOP OF INTERVAL (C83) . DEPTH TO BOTTOM OF INTERVAL (C84) . DIAMETER OF INTERVAL (C87) .

[5] MATERIAL TYPE (C86) [6] TYPE OF OPENING (C85) LENGTH OF OPENING (C89) . WIDTH OF OPENING (C88) .

FOOTNOTES:

[5] TYPE OF MATERIAL CODES FOR OPEN SECTIONS

A	B	C	D	E	F	G	H	I	J	K	L	M	N	P	Q	R	S	T	V	W	X	Y	Z	4	6
ABS	brass or bronze	concrete	ceramic	PTFE	fiber-glass	galv. iron	fiber-glass plastic	wrought iron	fiber-glass epoxy	PVC thread-ed	glass	other metal	PVC glued	PVC	FEP	stain-less steel	steel	tile	brick	mem-brane	steel carbon	steel galva-nized	other	stain-less 304	stain-less 316

[6] TYPE OF OPENINGS CODES

F	L	M	P	R	S	T	W	X	Z
fractured rock	louvered or shutter-type	mesh screen	perforated, porous or slotted	wire-wound screen	screen (unk.)	sand point screen	walled or shored	open hole	other

CONSTRUCTION MEASURING POINT DATA

RECORD TYPE (C766) MPNT RECORD SEQUENCE NO. (C728) BEGINNING DATE (C321) – – (month day year) ENDING DATE (C322) – –

M.P. HEIGHT (C323) . ALTITUDE OF MEASURING POINT (C325) ALTITUDE METHOD (C326) ALTITUDE ACCURACY (C327)

ALTITUDE DATUM (C328) M.P. REMARKS (C324)

RECORD READY FOR WEB (C857)

Y	C	P	L
ready to display	condi-tional	proprie-tary	local use only

CONSTRUCTION LIFT DATA

RECORD TYPE (C752) L I F T — RECORD SEQUENCE NO. (C254)

TYPE OF LIFT (C43): A air, B bucket, C centrifugal, J jet, P piston, R rotary, S submersible, T turbine, U unknown, X no lift, Z other

DATE RECORDED (C38) month – day – year

PUMP INTAKE DEPTH (C44)

TYPE OF POWER (C45): D diesel, E electric, G gasoline, H hand, L LP gas, N natural gas, S solar, W windmill, Z other

HORSEPOWER RATING (C46) .

MANUFACTURER (C48)

SERIAL NO. (C49)

POWER COMPANY (C50)

POWER COMPANY ACCOUNT NUMBER (C51)

POWER METER NUMBER (C52)

PUMP RATING (C53) (million gallons/units of fuel) .

ADDITIONAL LIFT (C255)

PERSON OR COMPANY MAINTAINING PUMP (C54)

RATED PUMP CAPACITY (gpm) (C268)

STANDBY POWER (C56) (see TYPE OF POWER)

HORSEPOWER OF STANDBY POWER SOURCE (C57) .

MISCELLANEOUS OWNER DATA

RECORD TYPE (C768) O W N R — RECORD SEQUENCE NO. (C718) — DATE OF OWNERSHIP (C159)

WU OWNER TYPE (C350): CP Corporation, GV Government, IN Individual, MI Military, OT Other, TG Tribal, WS Water Supplier

END DATE OF OWNERSHIP (C374)

OWNER'S NAME (C161)

EXAMPLES: JONES, RALPH A.
JONES CONSTRUCTION COMPANY

OWNER'S PHONE NUMBER (C351)

ACCESS TO OWNER'S NAME (C352): 0 Public Access, 1 Cooperator, 2 USGS Only, 3 District Only, 4 Proprietary

OWNER'S ADDRESS (LINE 1) (C353)

OWNER'S ADDRESS (LINE 2) (C354)

OWNER'S CITY NAME (C355)

STATE (C356)

OWNER'S ZIP CODE (C357) –

OWNER'S COUNTRY NAME (C358)

ACCESS TO OWNER'S PHONE/ADDRESS (C359): 0 Public Access, 1 Cooperator, 2 USGS Only, 3 District Only, 4 Proprietary

MISCELLANEOUS VISIT DATA

RECORD TYPE (C774) V I S T — RECORD SEQUENCE NO. (C737) — DATE OF VISIT (C187) month – day – year

NAME OF PERSON (C188)

MISCELLANEOUS OTHER ID DATA (2 sets shown)

RECORD TYPE (C770) O T I D

RECORD SEQUENCE NO. (C736)

OTHER ID (C190)

ASSIGNER (C191)

RECORD SEQUENCE NO. (C736)

OTHER ID (C190)

ASSIGNER (C191)

MISCELLANEOUS OTHER DATA

RECORD TYPE (C772) O T D T

RECORD SEQUENCE NO. (C312)

OTHER DATA TYPE (C181)

OTHER DATA LOCATION (C182) C Cooperator's Office, D District Office R Reporting Agency Z other

DATA FORMAT (C261) F files, M machine readable, P published, Z other

MISCELLANEOUS LOGS DATA (3 sets shown)

RECORD TYPE (C778) L O G S

RECORD SEQUENCE NO. (C739)

TYPE OF LOG (C199)

BEG NN NG DEPTH (C200) .

ENDING DEPTH (C201) .

SOURCE OF DATA (C202) A other gov't D driller G geol-ogist L logs M memory O owner R other reported S reporting agency Z other

DATA FORMAT (C225) F files M machine readable P published Z other

OTHER DATA LOCATION (C226) ____

RECORD TYPE (C778) L O G S

RECORD SEQUENCE NO. (C739)

TYPE OF LOG (C199)

BEG NNING DEPTH (C200) .

ENDING DEPTH (C201) .

SOURCE OF DATA (C202) A other gov't D driller G geol-ogist L logs M memory O owner R other reported S reporting agency Z other

DATA FORMAT (C225) F files M machine readable P published Z other

OTHER DATA LOCATION (C226) ____

RECORD TYPE (C778) L O G S

RECORD SEQUENCE NO. (C739)

TYPE OF LOG (C199)

BEGINNING DEPTH (C200) .

ENDING DEPTH (C201) .

SOURCE OF DATA (C202) A other gov't D driller G geol-ogist L logs M memory O owner R other reported S reporting agency Z other

DATA FORMAT (C225) F files M machine readable P published Z other

OTHER DATA LOCATION (C226) ____

ACOUSTIC LOG:
AS Sonic
AV Acoustic velocity
AW Acoustic waveform
AT Acous ic televiewer

CALIPER LOG:
CP Caliper
CS Caliper, single arm
CT Caliper, three arm
CM Caliper, multi arm
CA Caliper, acous ic

DRILLING LOG:
DT Drilling ime
DR Drillers
DG Geologists
DC Core

ELECTRIC LOG:
EE Electric
ER Single-point resistance
EP Spontaneous potential
EL Long-normal resis ivity
ES Short-normal resis ivity
EF Focused resistivity
ET Lateral resistivity
EN Microresistivity
EC Microresistivity, forused
EO Microresis ivity, lateral
ED Dipmeter

ELECTROMAGNETIC LOG:
MM Magnetic log
MS Magnetic susceptibiity log
MI Electromagne ic induc ion log
MD Electromagne ic dual induction log
MR Radar reflection image log
MV Radar direct-wave velocity log
MA Radar direct-wave amplitude log

FLUID LOG:
FC Fluid conductivity
FR Fluid resis ivity
FT Fluid temperature
FF Fluid differential temperature
FV Fluid velocity
FS Spinner flowmeter
FH Heat-pulse flowmeter
FE Electromagnetic flowmeter
FD Doppler flowmeter
FA Radioactive tracer
FY Dye tracer
FB Brine tracer

NUCLEAR LOG:
NG Gamma
NS Spectral gamma
NA Gamma-gamma
NN Neutron
NT Neutron activitation
NM Neuclear magnetic resonance

OPTICAL LOG:
OV Video
OF Fisheye video
OS Sidewall video
OT Optical televiewer

COMBINATION LOG:
ZF Gamma, fluid resistivity, temperature
ZI Gamma, electromagne ic induction
ZR Long/short normal resistivity
ZT Fluid resistivity, temperature
ZM Electromagnetic flowmeter, fluid resistivity, temperature
ZN Long/short normal resistivity, spontaneous poten ial
ZP Single-point resistance, spontaneous potential
ZE Gamma, long/short normal resistivity, spontaneous potential, single-point resistance, fluid resi ivity, temperature

WELL CONSTRUCTION LOG:
WC Casing collar
WD Borehold deviation

OTHER LOG:
OR Other

MISCELLANEOUS NETWORK DATA (3 types shown)

RECORD TYPE (C780) N E T W — RECORD SEQUENCE NO. (C730) — TYPE OF NETWORK (C706) Q W (water quality) — BEGINN NG YEAR (C115) — END NG YEAR (C116)

TYPE OF ANALYSIS (C120)

A	B	C	D	E	F	G	H	I	J	K	L	M	N	P	Z
physical proper-ties	common ions	trace elements	pesti-cides	nutri-ents	sanitary analysis	codes D&B	codes B&E	codes B&C	codes B&F	codes D&E	codes C,D&E	all or most	codes B&C& radio-active	codes B,C&A	other

SOURCE AGENCY (C117) — FREQUENCY OF COLLECTION (C118)[7] — ANALYZ NG AGENCY (C307) — PRIMARY NETWORK SITE (C257)[8] — SECONDARY NETWORK SITE (C708)[8]

RECORD TYPE (C780) N E T W — RECORD SEQUENCE NO. (C730) — TYPE OF NETWORK (C706) W L (water level) — BEGINN NG YEAR (C115) — END NG YEAR (C116)

SOURCE AGENCY (C117) — FREQUENCY OF COLLECTION (C118)[7] — PRIMARY NETWORK SITE (C257)[8] — SECONDARY NETWORK SITE (C708)[8]

RECORD TYPE (C780) N E T W — RECORD SEQUENCE NO. (C730) — TYPE OF NETWORK (C706) W D (pumpage or with-drawals) — BEGINN NG YEAR (C115) — END NG YEAR (C116)

SOURCE AGENCY (C117) — FREQUENCY OF COLLECTION (C118)[7] — METHOD OF COLLECTION (C133) — PR MARY NETWORK SITE (C257)[8] — SECONDARY NETWORK SITE (C708)[8]

C	E	M	U	Z
calcu-lated	esti-mated	meter-ed	un-known	other

FOOTNOTES:

[7] FREQUENCY OF COLLECTION CODES

A	B	C	D	F	I	M	O	Q	S	W	Z	2	3	4	5	X
annually	bi monthly	continu-ously	daily	semi-monthly	inter mittent	monthly	one-time only	quarter-ly	semi-annually	weekly	other	bi-annually	every 3 years	every 4 years	every 5 years	every 10 years

[8] NETWORK SITE CODES

1	2	3	4
national,	district,	project,	co-operator,

MISCELLANEOUS REMARKS DATA (4 types shown)

RECORD TYPE (C788) R M K S — RECORD SEQUENCE NO. (C311) — DATE OF REMARK (C184) month – day – year

REMARKS (C185)

Subsequent entries may be used to continue the remark. Miscellaneous remarks field is limited to 256 characters.

RECORD TYPE (C788) R M K S — RECORD SEQUENCE NO. (C311) — DATE OF REMARK (C184) month – day – year

REMARKS (C185)

Subsequent entries may be used to continue the remark. Miscellaneous remarks field is limited to 256 characters.

DISCHARGE DATA

RECORD SEQUENCE NO. (C147)

DATE DISCHARGE MEASURED (C148) month – day – year

TYPE OF DISCHARGE (C703) P pumped F flow

DISCHARGE (gpm) (C150) .

ACCURACY OF DISCHARGE MEASUREMENT (C310) E excellent (LT 2%), G good (2%-5%), F fair (5%-8%), P poor (GT 8%)

SOURCE OF DATA (C151) A other gov't, D driller, G geologist, L logs, M memory, O owner, R other reported, S reporting agency, Z other

METHOD OF DISCHARGE MEASUREMENT (C152) A acoustic meter, B bailer, C current meter, D Doppler meter, E estimated, F flume, M totaling meter, O orifice, P pitot-tube, R reported, T trajectory, U venturi meter, V volumetric meas, W weir, X unknown, Z other

PRODUCTION WATER LEVEL (C153) .

STATIC WATER LEVEL (C154) .

SOURCE OF DATA (C155) A other gov't, D driller, G geologist, L logs, M memory, O owner, R other reported, S reporting agency, Z other

METHOD OF WATER-LEVEL MEASUREMENT (C156) A airline, B recorder, C calibrated airline, D differential GP, E estimated, F transducer, G pressure gage, H calibrated press. gage, L geophysical logs, M manometer, N non-rec. gage, O observed, P acoustic pulse, R reported, S steel tape, T electric tape, V calibrated elec. tape, Z other

PUMP NG PERIOD (C157) .

SPECIFIC CAPACITY (C272) .

DRAWDOWN (C309) .

GEOHYDROLOGIC DATA

RECORD TYPE (C748) G E O H

RECORD SEQUENCE NO. (C721)

DEPTH TO TOP OF UNIT (C91) .

DEPTH TO BOTTOM OF UNIT (C92) .

UNIT IDENT FIER (C93)

LITHOLOGY (C96)

CONTRIBUTING UNIT (C304) P principal aquifer, Q aggregate of lithologic units, S secondary aquifer, N no contribution, U unknown

LITHOLOGIC MOD FIER (C97)

GEOHYDROLOGIC AQUIFER DATA

RECORD TYPE (C750) A Q F R

RECORD SEQUENCE NO. (C742)

SEQUENCE NO. OF PARENT RECORD (C256)

DATE (C95) month – day – year

STATIC WATER LEVEL (C126) .

CONTR BUTION (C132)

SITE LOCATION SKETCH AND DIRECTIONS

Township ___________ Range ___________

Section #___________

GWPD 11—Measuring well depth by use of a graduated steel tape

VERSION: 2010.1

PURPOSE: To measure the total depth of a well below land-surface datum by using a weighted graduated steel tape.

Materials and Instruments

1. A steel tape graduated in feet, tenths and hundredths of feet. A break-away weight should be attached to a ring on the end of the tape with wire strong enough to hold the weight, but not as strong as the tape, so that if the weight becomes lodged in the well the tape can still be pulled free. The weight should be made of brass, stainless steel, or iron. A lead weight should not be used. The weight should be heavy enough to amplify the weight-transfer sensation when the bottom of the well is struck.
2. Clean rag
3. Cleaning supplies for water-level tapes as described in the National Field Manual (Wilde, 2004)
4. Two wrenches with adjustable jaws or other tools for removing well cap
5. Key for well access
6. Pencil or pen, blue or black ink. Strikethrough, date, and initial errors; no erasures
7. Field notebook
8. Groundwater Site Inventory (GWSI) System, Groundwater Site Schedule Form 9-1904-A

Data Accuracy and Limitations

1. A graduated steel tape is commonly accurate to 0.01 foot. Accuracy of well-depth measurement decreases with increasing depth.
2. The steel tape should be calibrated against another acceptable steel tape. An acceptable steel tape is one that is maintained in the office for use only for calibrating steel and electric tapes.
3. Corrections are necessary for measurements made in angled well casings.
4. When measuring well depth in deep wells, tape expansion and stretch is an additional consideration (Garber and Koopman, 1968).

Advantages

1. The weighted graduated steel tape is considered to be the most accurate method of measuring well depth.
2. Easy to use.

Disadvantages

1. Not recommended for measuring the depth of wells that are being pumped.

Assumptions

1. An established measuring point (MP) exists. See GWPD 3 for technical procedures on establishing an MP.
2. The MP is clearly marked and described.
3. The steel tape has been calibrated.
4. The well is free of obstructions that could affect the plumbness of the steel tape and cause errors in the measurement.

Instructions

1. Measure from the zero point on the tape to the bottom of the weight. Record this number in the field notebook as the length of the weight interval.
2. Lower the weight and tape into the well until the weight reaches the bottom of the well and the tape slackens.
3. Partially withdraw the tape from the well until the weight is standing in a vertical position, but still touching the bottom of the well. A slight jerking motion will be felt as the weight moves from the horizontal to the vertical position.
4. Repeat step 3 several times by lowering and withdrawing the tape to obtain a consistent reading.
5. Record the tape reading held at the MP.
6. Withdraw the tape from the well 1 to 2 feet, so that the weight will hang freely above the bottom of the well. Repeat steps 2–4 until two consistent depth readings are obtained.
7. Calculate total well depth below land-surface datum (LSD) as follows:

Tape reading held at the MP	84.30 feet
Length of the weight interval	+1.20 feet
Total well depth below MP	85.50 feet
MP correction	–3.40 feet
Total well depth below LSD	82.10 feet

8. After completing the well-depth measurement, disinfect and rinse that part of the tape that was submerged below the water surface, as described in the National Field Manual (Wilde, 2004). This will reduce the possibility of contamination of other wells from the tape.

Data Recording

Data are recorded in a field notebook. Well-depth data are recorded in the groundwater site data section of the GWSI Groundwater Site Schedule (fig. 1, Form 9-1904-A). Recommended precision is depth dependent and should be shown in field C28 on Form 9-1904-A (fig. 1).

References

Cunningham, W.L., and Schalk, C.W., comps., 2011, Groundwater technical procedures of the U.S. Geological Survey, GWPD 3—Establishing a permanent measuring point and other reference marks: U.S. Geological Survey Techniques and Methods 1–A1, 13 p.

Garber, M.S., and Koopman, F.C., 1968, Methods of measuring water levels in deep wells: U.S. Geological Survey Techniques of Water-Resources Investigations, book 8, chap. A1, 23 p.

Hoopes, B.C., ed., 2004, User's manual for the National Water Information System of the U.S. Geological Survey, Ground-Water Site-Inventory System (version 4.4): U.S. Geological Survey Open-File Report 2005–1251, 274 p.

Katz, B.G., and Jelinski, J.C., 1999, Replacement materials for lead weights used in measuring ground-water levels: U.S. Geological Survey Open-File Report 99–52, 13 p.

Wilde, F.D., ed., 2004, Cleaning of equipment for water sampling (version 2.0): U.S. Geological Survey Techniques of Water-Resources Investigations, book 9, chap. A3, section 3.3.8., p. 50–53, accessed May 17, 2010, at *http://pubs.water.usgs.gov/twri9A3/*.

FORM NO. 9-1904-A
Revised Sept 2009, NWIS 4.9

File Code ______
Date ______

Coded by ______
Checked by ______
Entered by ______

U.S DEPT. OF THE INTERIOR
GEOLOGICAL SURVEY

GROUNDWATER SITE SCHEDULE
General Site Data

AGENCY CODE (C4) USGS | SITE ID (C1) | PROJECT (C5)

STATION NAME (C12/900)

SITE TYPE[1] (C802) Primary - Secondary | DISTRICT (C6) | COUNTRY (C41) | STATE (C7)

COUNTY or TOWN (C8) ______ County code

LATITUDE (C9) . | LONGITUDE (C10) . | LAT/LONG ACCURACY (C11) H (Hndrth sec.) 1 (tenth sec.) 5 (half sec.) S (sec.) R (3 sec.) F (5 sec.) T (10 sec.) M (min.) U (Un-known)

LAT/LONG METHOD (C35) C (land net) D (DGPS) G (GPS) L (LORAN) M (map) N (inter-polated digital map) R (reported) S (survey) U (un-known)

LAT/LONG DATUM (C36) NAD27 (North American Datum of 1927) NAD83 (North American Datum of 1983)

ALTITUDE (C16) .

ALTITUDE ACCURACY (C18)

ALTITUDE METHOD (C17) A (altimeter) D (DGPS) G (GPS) I (IfSAR) J (LIDAR) L (Level) M (map) N (DEM) R (re-ported) U (un-known)

ALTITUDE DATUM (C22) NGVD29 (National Geodetic Vertical Datum of 1929) NAVD88 (North American Vertical Datum of 1988)

LAND NET (C13) ¼ ¼ ¼ S section T township range merid

TOPO-GRAPHIC SETTING (C19) A (alluvial fan) B (playa) C (stream channel) D (depres-sion) E (dunes) F (flat) G (flood-plain) H (hill-top) K (sink-hole) L (lake or swamp) M (mangrove swamp) O (off-shore) P (pedi-ment) S (hill-side) T (ter-race) U (undu-lating) V (valley flat) W (upland draw)

HYDROLOGIC UNIT CODE (C20) | DRAINAGE BASIN CODE (C801) | STANDARD TIME ZONE (C813) | DAYLIGHT SAVINGS TIME FLAG (C814) Y OR **N**

MAP NAME (C14) | MAP SCALE (C15)

AGENCY USE (C803) A (active no/na) D (discon-tinued) I (inactive site) L (active written) M (active oral) O (inventory site) R (remediated)

2 NATIONAL WATER-USE (C39)

DATA TYPE (C804)
Place an 'A' (active), an 'I' (inactive), or an 'O' (inventory) in the appropriate box

WL cont | WL int | QW cont | QW int | PR cont | PR int | EV cont | EV int | wind vel. | tide cont | tide int | sed. con | sed. ps | peak flow | low flow | state water use

INSTRUMENTS (C805)
(Place a "Y" in the appropriate box):

digital rec-order | graphic rec-order | tele-metry land line | tele-metry radio | tele-metry satellite | AHDAS | crest-stage gage | tide gage | deflec-tion meter | bubble gage | stilling well | CR type recorder | weigh-ing rain gage | tipping bucket rain gage | acoustic velocity meter | electro-magnetic flowmeter | pressure transducer

DATE INVENTORIED (C711) month - day - year

RECORD READY FOR WEB (C32) Y (ready to display) C (condi-tional) P (proprie-tary) L (local use only)

REMARKS (C806)

FOOTNOTES

1SITE TYPE (C802)

GL	Glacier	OC	Ocean	GW	Well	SB	Subsurface
WE	Wetland	OC-CO	Coastal	GW-CR	Collector or Ranney type well	SB-CV	Cave
AT	Atmosphere	LK	Lake, Reservoir, Impoundment	GW-EX	Extensometer well	SB-GWD	Groundwater drain
ES	Estuary			GW-HZ	Hyporheic -zone well	SB-TSM	Tunnel, shaft, or mine
LA	Land	SP	Spring	GW-IW	Interconnected wells	SB-UZ	Unsaturated zone
LA-EX	Excavation	ST	Stream	GW-TH	Test hole not completed as a well		
LA-OU	Outcrop	ST-CA	Canal	GW-MW	Multiple wells		
LA-SNK	Sinkhole	ST-DCH	Ditch				
LA-SH	Soil hole	ST-TS	Tidal stream				
LA-SR	Shore	FA-WIW	Waste-Injection well				

2 WS (water supply) | DO (domestic) | CO (commer-cial) | IN (industrial) | IR (irrigation) | MI (mining) | LV (livestock) | PH (power hydro-electric) | ST (waste water treatment) | RM (remedia-tion) | TE (thermo-electric power) | AQ (aqua-culture)

C22 Other (see manual for codes)
C36 Other (see manual for codes)
C39 is mandatory for all sites having data in SWUDS.

Figure 1. Groundwater Site Schedule, Form 9-1904-A.

GENERAL SITE DATA

DATA RELIABILITY (C3): C (field checked), L (poor location), M (minimal data), U (unchecked)

DATE OF F RST CONSTRUCTION (C21): [|] – [|] – [| | |] (month, day, year)

USE OF SITE (C23): A (anode), C (standby emer. supply), D (drain), E (geo-thermal), G (seismic), H (heat reservoir), M (mine), O (obser-vation), P (oil or gas), R (recharge), S (repres-surize), T (test), U (unused), V (with-drawal/return), W (with-drawal), X (waste), Z (des-troyed)

SECOND-ARY USE OF SITE (C301) (See use of site) [ ]   TERTIARY USE OF SITE (C302) (See use of site) [ ]

USE OF WATER (C24): A (air cond.), B (bottling), C (comm-ercial), D (de-water), E (power), F (fire), H (domes-tic), I (irri-gation), J (indus-trial (cooling)), K (mining), M (medi-cinal), N (indus-trial), P (public supply), Q (aqua-culture), R (recrea-tions), S (stock), T (insti-tutional), U (unused), Y (desalin-ation), Z (other)

SECOND-ARY USE OF WATER (C25) (see use of water) [ ]   TERTIARY USE OF WATER (C26) (see use of water) [ ]

AQU FER TYPE (C713): U (unconfined single), N (unconfined multiple), C (confined single), M (confined multiple), X (mixed)

PR MARY AQUIFER (C714) [| | | | | | |]   NATIONAL AQU FER (C715) [| | | | | | | | |]

HOLE DEPTH (C27) [| | | |] . [|]   WELL DEPTH (C28) [| | | |] . [|]

SOURCE OF DEPTH DATA (C29): A (other gov't), D (driller), G (geol-ogist), L (logs), M (memory), O (owner), R (other reported), S (reporting agency), Z (other)

WATER-LEVEL DATA

DATE WATER-LEVEL MEASURED (C235) [|] – [|] – [| | |] (month, day, year)

T ME (C709) [| | |]

WATER-LEVEL TYPE CODE (C243): L (land surface), M (meas. pt.), S (vertical datum)

WATER LEVEL (C237/241/242) [| | |] . [|]

MP SEQUENCE NO. (C248) (Mandatory if WL type=M) [| |]

WATER-LEVEL DATUM (C245) (Mandatory if WL type=S): NGVD29 (National Geodetic Vertical Datum 0f 1929), NAVD88 (North American Vertical Datum 0f 1988), [| | | | | | | | | |] Other (See manual for codes)

SITE STATUS FOR WATER LEVEL (C238): A (atmos. pressure), B (tide stage), C (ice), D (dry), E (recently flowing), F (flowing), G (nearby flowing), H (nearby recently flowing), I (injector site), J (injector site monitor), M (plugged), N (measure-ment discontinued), O (obstruc-tion), P (pumping), R (recently pumped), S (nearby pumping), T (nearby recently pumped), V (foreign sub-stance), W (well des-troyed), X (affected by surface water), Z (other)

METHOD OF WATER-LEVEL MEASUREMENT(C239): A (airline), B (analog), C (calibrated airline), D (differ-ential GPS), E (esti-mated), F (trans-ducer), G (pressure gage), H (calibrated press. gage), L (geophysi-cal logs), M (mano-meter), N (non-rec. gage), O (observed), P (acoustic pulse), R (reported), S (steel tape), T (electric tape), V (calibrated elec. tape), Z (other)

WATER-LEVEL ACCURACY (C276): 0 (foot), 1 (tenth), 2 (hun-dredth), 9 (not to nearest foot)

SOURCE OF WATER-LEVEL DATA (C244): A (other gov't), D (driller's log), G (geol-ogist), L (geophysi-cal logs), M (memory), O (owner), R (other reported), S (reporting agency), Z (other)

PERSON MAKING MEASUREMENT (C246) (WATER LEVEL PARTY) [| | | | |]

MEASURING AGENCY (C247) (SOURCE) [| | | |]

EQUIP D (C249) (20 char) ____________

REMARKS (C267) (256 char) ____________

RECORD READY FOR WEB (C858): Y (ready to display), C (condi-tional), P (proprie-tary), L (local use only)

CONSTRUCTION DATA

RECORD TYPE (C754): C O N S

RECORD SEQUENCE NO. (C723) [| |]

DATE OF COMPLETED CONSTRUCTION (C60) [|] – [|] – [| | |] (month, day, year)

NAME OF CONTRACTOR (C63) [| | | | | | | | | | | |]

SOURCE OF DATA (C64): A (other gov't), D (driller), G (geol-ogist), L (logs), M (memory), O (owner), R (other reported), S (reporting agency), Z (other)

METHOD OF CONSTRUCTION (C65): A (air-rotary), B (bored or augered), C (cable tool), D (dug), H (hydraulic rotary), J (jetted), P (air per-cussion), R (reverse rotary), S (sonic), T (trenching), V (driven), W (drive wash), Z (other)

TYPE OF FINISH (C66): C (porous concrete), F (gravel w/perf.), G (gravel screen), H (horiz. gallery), O (open end), P (perf or slotted), S (screen), T (sand point), W (walled), X (open hole), Z (other)

TYPE OF SEAL (C67): B (bentonite), C (clay), G (cement grout), N (none), Z (other)

BOTTOM OF SEAL (C68) [| | |]

METHOD OF DEVELOPMENT (C69): A (air-lift pump), B (bailed), C (compres-sed air), J (jetted), N (none), P (pumped), S (surged), Z (other)

HOURS OF DEVELOPMENT (C70) [| |]

SPECIAL TREATMENT (C71): C (chem-icals), D (dry ice), E (explo-sives), F (defloc-culent), H (hydro-frac-turing), M (mech-anical), Z (other)

2 - Groundwater Site Schedule

CONSTRUCTION HOLE DATA (3 sets shown)

RECORD TYPE (C756) HOLE RECORD SEQUENCE NO. (C724) SEQUENCE NO. OF PARENT RECORD (C59)

DEPTH TO TOP OF INTERVAL (C73) . DEPTH TO BOTTOM OF NTERVAL (C74) . DIAMETER OF INTERVAL (C75) .

RECORD SEQUENCE NO. (C724)

DEPTH TO TOP OF INTERVAL (C73) . DEPTH TO BOTTOM OF NTERVAL (C74) . DIAMETER OF INTERVAL (C75) .

RECORD SEQUENCE NO. (C724)

DEPTH TO TOP OF INTERVAL (C73) . DEPTH TO BOTTOM OF NTERVAL (C74) . DIAMETER OF INTERVAL (C75) .

CONSTRUCTION CASING DATA (4 sets shown)

RECORD TYPE (C758) CSNG RECORD SEQUENCE NO. (C725) SEQUENCE NO. OF PARENT RECORD (C59)

DEPTH TO TOP OF CAS NG (C77) . DEPTH TO BOTTOM OF CASING (C78) . DIAMETER OF CASING (C79) .

4 CASING MATERIAL (C80) CASING THICKNESS (C81) .

RECORD SEQUENCE NO. (C725) SEQUENCE NO. OF PARENT RECORD (C59)

DEPTH TO TOP OF CASING (C77) . DEPTH TO BOTTOM OF CASING (C78) . DIAMETER OF CASING (C79) .

4 CASING MATERIAL (C80) CASING THICKNESS (C81) .

RECORD SEQUENCE NO. (C725) SEQUENCE NO. OF PARENT RECORD (C59)

DEPTH TO TOP OF CAS NG (C77) . DEPTH TO BOTTOM OF CASING (C78) . DIAMETER OF CASING (C79) .

4 CAS NG MATERIAL (C80) CAS NG THICKNESS (C81) .

RECORD SEQUENCE NO. (C725) SEQUENCE NO. OF PARENT RECORD (C59)

DEPTH TO TOP OF CAS NG (C77) . DEPTH TO BOTTOM OF CASING (C78) . DIAMETER OF CASING (C79) .

4 CAS NG MATERIAL (C80) CAS NG THICKNESS (C81) .

FOOTNOTE:

4 CAS NG MATERIAL CODES	A	B	C	D	E	F	G	H	I	J	K	L	M	N	P	Q	R	S	T	U	V	W	X	Y	Z	4	6
	abs	brick	concrete	copper	PTFE	Fiber-glass	galv. iron	Fiber-glass plastic	wrought iron	Fiber-glass epoxy	PVC thread-ed	glass	other metal	PVC glued	PVC or plastic	FEP	rock or stone	steel	tile	coated steel	stain-less steel	wood	steel carbon	steel galva-nized	other mat.	stain-less 304	stain-less 316

CONSTRUCTION OPENINGS DATA (3 sets shown)

RECORD TYPE (C760) OPEN   RECORD SEQUENCE NO. (C726)   SEQUENCE NO. OF PARENT RECORD (C59)

DEPTH TO TOP OF NTERVAL (C83)   DEPTH TO BOTTOM OF INTERVAL (C84)   DIAMETER OF INTERVAL (C87)

[5]MATERIAL TYPE (C86)   [6]TYPE OF OPEN NG (C85)   LENGTH OF OPENING (C89)   WIDTH OF OPENING (C88)

RECORD SEQUENCE NO. (C726)

DEPTH TO TOP OF INTERVAL (C83)   DEPTH TO BOTTOM OF INTERVAL (C84)   DIAMETER OF NTERVAL (C87)

[5]MATERIAL TYPE (C86)   [6]TYPE OF OPEN NG (C85)   LENGTH OF OPENING (C89)   WIDTH OF OPENING (C88)

RECORD SEQUENCE NO. (C726)

DEPTH TO TOP OF INTERVAL (C83)   DEPTH TO BOTTOM OF INTERVAL (C84)   DIAMETER OF NTERVAL (C87)

[5]MATERIAL TYPE (C86)   [6]TYPE OF OPEN NG (C85)   LENGTH OF OPENING (C89)   WIDTH OF OPENING (C88)

FOOTNOTES:

[5] TYPE OF MATERIAL CODES FOR OPEN SECTIONS

A	B	C	D	E	F	G	H	I	J	K	L	M	N	P	Q	R	S	T	V	W	X	Y	Z	4	6
ABS	brass or bronze	concrete	ceramic	PTFE	fiber-glass	galv. iron	fiber-glass plastic	wrought iron	fiber-glass epoxy	PVC thread-ed	glass	other metal	PVC glued	PVC	FEP	stain-less steel	steel	tile	brick	mem-brane	steel carbon	steel galva-nized	other	stain-less 304	stain-less 316

[6] TYPE OF OPENINGS CODES

F	L	M	P	R	S	T	W	X	Z
fractured rock	louvered or shutter-type	mesh screen	perforated, porous or slotted	wire-wound screen	screen (unk.)	sand point screen	walled or shored	open hole	other

CONSTRUCTION MEASURING POINT DATA

RECORD TYPE (C766) MPNT   RECORD SEQUENCE NO. (C728)   BEGINNING DATE (C321) month – day – year   ENDING DATE (C322)

M.P. HEIGHT (C323)   ALTITUDE OF MEASURING PO NT (C325)   ALTITUDE METHOD (C326)   ALTITUDE ACCURACY (C327)

ALTITUDE DATUM (C328)   M.P. REMARKS (C324)

RECORD READY FOR WEB (C857)

Y	C	P	L
ready to display	condi-tional	proprie-tary	local use only

CONSTRUCTION LIFT DATA

RECORD TYPE (C752) L I F T — RECORD SEQUENCE NO. (C254) — TYPE OF LIFT (C43): A air, B bucket, C centri-fugal, J jet, P piston, R rotary, S submer-sible, T turbine, U un-known, X no lift, Z other

DATE RECORDED (C38) month – day – year — PUMP INTAKE DEPTH (C44) — TYPE OF POWER (C45): D diesel, E electric, G gaso-line, H hand, L LP gas, N natural gas, S solar, W windmill, Z other

HORSE-POWER RATING (C46) . — MANUFACTURER (C48) — SERIAL NO. (C49)

POWER COMPANY (C50) — POWER COMPANY ACCOUNT NUMBER (C51)

POWER METER NUMBER (C52) — PUMP RATING (C53) (million gallons/units of fuel) . — ADDITIONAL LIFT (C255)

PERSON OR COMPANY MAINTAINING PUMP (C54) — RATED PUMP CAPACITY (gpm) (C268) — STANDBY POWER (C56) (see TYPE OF POWER)

HORSEPOWER OF STANDBY POWER SOURCE (C57) .

MISCELLANEOUS OWNER DATA

RECORD TYPE (C768) O W N R — RECORD SEQUENCE NO. (C718) — DATE OF OWNERSH P (C159) – –

WU OWNER TYPE (C350): CP Corporation, GV Govern-ment, IN Individual, MI Military, OT Other, TG Tribal, WS Water Supplier — END DATE OF OWNERSHIP (C374) – –

OWNER'S NAME (C161)

EXAMPLES: JONES, RALPH A.
JONES CONSTRUCTION COMPANY

OWNER'S PHONE NUMBER (C351) — ACCESS TO OWNER'S NAME (C352): 0 Public Access, 1 Coop-erator, 2 USGS Only, 3 District Only, 4 Proprietary

OWNER'S ADDRESS (LINE 1) (C353)

OWNER'S ADDRESS (LINE 2) (C354)

OWNER'S CITY NAME (C355)

STATE (C356) — OWNER'S Z P CODE (C357) –

OWNER'S COUNTRY NAME (C358)

ACCESS TO OWNER'S PHONE/ADDRESS (C359): 0 Public Access, 1 Coop-erator, 2 USGS Only, 3 District Only, 4 Proprietary

MISCELLANEOUS VISIT DATA

RECORD TYPE (C774) V I S T — RECORD SEQUENCE NO. (C737) — DATE OF VISIT (C187) month – day – year

NAME OF PERSON (C188)

MISCELLANEOUS OTHER ID DATA (2 sets shown)

RECORD TYPE (C770) O|T|I|D RECORD SEQUENCE NO. (C736) OTHER ID (C190)

ASSIGNER (C191)

RECORD SEQUENCE NO. (C736) OTHER ID (C190)

ASSIGNER (C191)

MISCELLANEOUS OTHER DATA

RECORD TYPE (C772) O|T|D|T RECORD SEQUENCE NO. (C312)

OTHER DATA TYPE (C181)

OTHER DATA LOCATION (C182) C Cooperator's Office, D District Office R Reporting Agency Z other

DATA FORMAT (C261) F files, M machine readable, P published, Z other

MISCELLANEOUS LOGS DATA (3 sets shown)

RECORD TYPE (C778) L|O|G|S RECORD SEQUENCE NO. (C739) TYPE OF LOG (C199)

BEG NN NG DEPTH (C200) . ENDING DEPTH (C201) . SOURCE OF DATA (C202) A other gov't D driller G geol-ogist L logs M memory O owner R other reported S reporting agency Z other

DATA FORMAT (C225) F files M machine readable P published Z other OTHER DATA LOCATION (C226)

RECORD TYPE (C778) L|O|G|S RECORD SEQUENCE NO. (C739) TYPE OF LOG (C199)

BEG NNING DEPTH (C200) . ENDING DEPTH (C201) . SOURCE OF DATA (C202) A other gov't D driller G geol-ogist L logs M memory O owner R other reported S reporting agency Z other

DATA FORMAT (C225) F files M machine readable P published Z other OTHER DATA LOCATION (C226)

RECORD TYPE (C778) L|O|G|S RECORD SEQUENCE NO. (C739) TYPE OF LOG (C199)

BEGINNING DEPTH (C200) . ENDING DEPTH (C201) . SOURCE OF DATA (C202) A other gov't D driller G geol-ogist L logs M memory O owner R other reported S reporting agency Z other

DATA FORMAT (C225) F files M machine readable P published Z other OTHER DATA LOCATION (C226)

ACOUSTIC LOG:
AS Sonic
AV Acoustic velocity
AW Acoustic waveform
AT Acous ic televiewer

CALIPER LOG:
CP Caliper
CS Caliper, single arm
CT Caliper, three arm
CM Caliper, multi arm
CA Caliper, acous ic

DRILLING LOG:
DT Drilling ime
DR Drillers
DG Geologists
DC Core

ELECTRIC LOG:
EE Electric
ER Single-point resistance
EP Spontaneous potential
EL Long-normal resistivity
ES Short-normal resis ivity
EF Focused resistivity
ET Lateral resistivity
EN Microresistivity
EC Microresistivity, forused
EO Microresis ivity, lateral
ED Dipmeter

ELECTROMAGNETIC LOG:
MM Magnetic log
MS Magnetic susceptibiity log
MI Electromagne ic induc ion log
MD Electromagne ic dual induction log
MR Radar reflection image log
MV Radar direct-wave velocity log
MA Radar direct-wave amplitude log

FLUID LOG:
FC Fluid conductivity
FR Fluid resistivity
FT Fluid temperature
FF Fluid differen ial temperature
FV Fluid velocity
FS Spinner flowmeter
FH Heat-pulse flowmeter
FE Electromagnetic flowmeter
FD Doppler flowmeter
FA Radioactive tracer
FY Dye tracer
FB Brine tracer

NUCLEAR LOG:
NG Gamma
NS Spectral gamma
NA Gamma-gamma
NN Neutron
NT Neutron activitation
NM Neuclear magnetic resonance

OPTICAL LOG:
OV Video
OF Fisheye video
OS Sidewall video
OT Optical televiewer

COMBINATION LOG:
ZF Gamma, fluid resistivity, temperature
ZI Gamma, electromagne ic induction
ZR Long/short normal resistivity
ZT Fluid resis ivity, temperature
ZM Electromagnetic flowmeter, fluid resistivity, temperature
ZN Long/short normal resistivity, spontaneous poten ial
ZP Single-point resistance, spontaneous potential
ZE Gamma, long/short normal resistivity, spontaneous potential, single-point resistance, fluid resi ivity, temperature

WELL CONSTRUCTION LOG:
WC Casing collar
WD Borehold deviation

OTHER LOG:
OR Other

MISCELLANEOUS NETWORK DATA (3 types shown)

RECORD TYPE (C780) N E T W | RECORD SEQUENCE NO. (C730) | TYPE OF NETWORK (C706) Q W water quality | BEGINNING YEAR (C115) | ENDING YEAR (C116)

TYPE OF ANALYSIS (C120)

A	B	C	D	E	F	G	H	I	J	K	L	M	N	P	Z
physical proper-ties	common ions	trace elements	pesti-cides	nutri-ents	sanitary analysis	codes D&B	codes B&E	codes B&C	codes B&F	codes D&E	codes C,D&E	all or most	codes B&C& radio-active	codes B,C&A	other

SOURCE AGENCY (C117) | 7 FREQUENCY OF COLLECTION (C118) | ANALYZING AGENCY (C307) | 8 PRIMARY NETWORK SITE (C257) | 8 SECONDARY NETWORK SITE (C708)

RECORD TYPE (C780) N E T W | RECORD SEQUENCE NO. (C730) | TYPE OF NETWORK (C706) W L water level | BEGINNING YEAR (C115) | ENDING YEAR (C116)

SOURCE AGENCY (C117) | 7 FREQUENCY OF COLLECTION (C118) | 8 PRIMARY NETWORK SITE (C257) | 8 SECONDARY NETWORK SITE (C708)

RECORD TYPE (C780) N E T W | RECORD SEQUENCE NO. (C730) | TYPE OF NETWORK (C706) W D pumpage or with-drawals | BEGINNING YEAR (C115) | ENDING YEAR (C116)

SOURCE AGENCY (C117) | 7 FREQUENCY OF COLLECTION (C118) | METHOD OF COLLECTION (C133) | 8 PRIMARY NETWORK SITE (C257) | 8 SECONDARY NETWORK SITE (C708)

C	E	M	U	Z
calcu-lated	esti-mated	meter-ed	un-known	other

FOOTNOTES:

7 FREQUENCY OF COLLECTION CODES

A	B	C	D	F	I	M	O	Q	S	W	Z	2	3	4	5	X
annually	bi monthly	continu-ously	daily	semi-monthly	inter mittent	monthly	one-time only	quarter-ly	semi-annually	weekly	other	bi-annually	every 3 years	every 4 years	every 5 years	every 10 years

8 NETWORK SITE CODES

1	2	3	4
national,	district,	project,	co-operator,

MISCELLANEOUS REMARKS DATA (4 types shown)

RECORD TYPE (C788) R M K S | RECORD SEQUENCE NO. (C311) | DATE OF REMARK (C184) month – day – year

REMARKS (C185)

Subsequent entries may be used to continue the remark. Miscellaneous remarks field is limited to 256 characters.

RECORD TYPE (C788) R M K S | RECORD SEQUENCE NO. (C311) | DATE OF REMARK (C184) month – day – year

REMARKS (C185)

Subsequent entries may be used to continue the remark. Miscellaneous remarks field is limited to 256 characters.

DISCHARGE DATA

RECORD SEQUENCE NO. (C147)

DATE DISCHARGE MEASURED (C148) month – day – year

TYPE OF DISCHARGE (C703) P pumped F flow

DISCHARGE (gpm) (C150) .

ACCURACY OF DISCHARGE MEASUREMENT (C310) E excellent (LT 2%), G good (2%-5%), F fair (5%-8%), P poor (GT 8%)

SOURCE OF DATA (C151) A other gov't, D driller, G geologist, L logs, M memory, O owner, R other reported, S reporting agency, Z other

METHOD OF DISCHARGE MEASUREMENT (C152) A acoustic meter, B bailer, C current meter, D Doppler meter, E estimated, F flume, M totaling meter, O orifice, P pitot-tube, R reported, T trajectory, U venturi meter, V volumetric meas, W weir, X unknown, Z other

PRODUCTION WATER LEVEL (C153) .

STATIC WATER LEVEL (C154) .

SOURCE OF DATA (C155) A other gov't, D driller, G geologist, L logs, M memory, O owner, R other reported, S reporting agency, Z other

METHOD OF WATER-LEVEL MEASUREMENT (C156) A airline, B recorder, C calibrated airline, D differential GP, E estimated, F transducer, G pressure gage, H calibrated press. gage, L geophysical logs, M manometer, N non-rec. gage, O observed, P acoustic pulse, R reported, S steel tape, T electric tape, V calibrated elec. tape, Z other

PUMPING PERIOD (C157) .

SPECIFIC CAPACITY (C272) .

DRAWDOWN (C309) .

GEOHYDROLOGIC DATA

RECORD TYPE (C748) GEOH

RECORD SEQUENCE NO. (C721)

DEPTH TO TOP OF UNIT (C91) .

DEPTH TO BOTTOM OF UNIT (C92) .

UNIT IDENTIFIER (C93)

LITHOLOGY (C96)

CONTRIBUTING UNIT (C304) P principal aquifer, Q aggregate of lithologic units, S secondary aquifer, N no contribution, U unknown

LITHOLOGIC MOD FIER (C97)

GEOHYDROLOGIC AQUIFER DATA

RECORD TYPE (C750) AQFR

RECORD SEQUENCE NO. (C742)

SEQUENCE NO. OF PARENT RECORD (C256)

DATE (C95) month – day – year

STATIC WATER LEVEL (C126) .

CONTRIBUTION (C132)

SITE LOCATION SKETCH AND DIRECTIONS

Township ___________ Range ___________

Section #___________

8 - Groundwater Site Schedule

GWPD 12—Measuring water levels in a flowing well

VERSION: 2010.1

PURPOSE: To measure low-pressure or high-pressure hydraulic head in flowing wells.

Materials and Instruments

1. Low-pressure head measurement
   - Short length of transparent plastic tubing
   - Hose clamps
   - Measuring scale
2. High-pressure head measurement
   - Flexible hose with a 3-way valve
   - Hose clamps
   - Altitude or pressure gauge with proper pressure range, and spare gauges
   - Small open end wrench
   - Soil-pipe test plug, also known as a sanitary seal, is a length of small-diameter pipe, generally 0.75 inch, surrounded by a rubber packer. The packer can be expanded by an attached wingnut to fit tightly against the inside of the well casing or discharge pipe. Soil-pipe test plugs are available from most plumbing-supply stores in 2- to 10-inch diameter sizes. The small-diameter pipe is threaded so that it can be attached to a valve, hose, or pressure gauge.
3. Pencil or pen, blue or black ink. Strikethrough, date, and initial errors; no erasures
4. Calibration and maintenance logbook
5. Water-level measurement field form

Data Accuracy and Limitations

1. Low-pressure head measurements are most feasible with heads less than 6 feet above land surface.
2. With care and experience, low-pressure head measurements can be measured to an accuracy of 0.1 foot.
3. Accuracy is a function of calibration, maintenance, and the quality and range of the pressure gauge. High-pressure head measurements using a pressure gauge can be as accurate as 0.1 foot, but may only be accurate to 1 foot or more, depending on the gauge accuracy and range.
4. A pressure gauge is the most accurate in the middle third of the gauge's range. Never let the well pressure exceed the altitude/pressure gauge limits.
5. Never connect a gauge to a well that uses a booster pump in the system, because the pump could start automatically and the resulting pressure surge may ruin the gauge.
6. Closing or opening a valve or test plug in a flowing well should be done gradually. If pressure is applied or released suddenly, the well could be permanently damaged by the "water-hammer effect" by caving of the aquifer material, breakage of the well casing, or damage to the distribution lines or gauges. To reduce the possibility of water-hammer effect, a pressure-snubber should be installed ahead of the altitude/pressure gauge.
7. Ideally, all flow from the well should be shut down so that a static water-level measurement can be made. However, because of well owner objections or system leaks, this is not always possible. If the well does not have a shut-down valve, it can be shut-in by temporarily installing a soil-pipe test plug on the well or discharge line.
8. If a well has to be shut down, the time required to reach static pressure after shut-in may range from hours to days. Since it may be impractical or impossible to reach true static conditions, record the shut-in time for each gauge reading. During return visits to a particular well, it is desirable to duplicate the previously used shut-in time before making an altitude/pressure-gauge reading.

Advantages

1. Low-pressure head measurement
   - Simpler, faster, safer, and more accurate than the high-pressure head method.
2. High-pressure head measurement
   - Can be used on wells with heads greater than 5 to 6 feet above land surface.

Disadvantages

1. Low-pressure head measurement
   - Impractical for wells with heads greater than 5 to 6 feet above land surface.
2. High-pressure head measurement
   - More complex, slower, less accurate, and more dangerous to make than low-pressure head measurements.
   - Pressure gauges are delicate, easily broken, and subject to erroneous readings if dropped or mistreated.
   - Difficult to calibrate.

Assumptions

1. An established measuring point (MP) exists. See GWPD 3 for technical procedures on establishing an MP.
2. Pressure gauges have been calibrated with a dead-weight tester.
3. A logbook containing all calibration and maintenance records is available for each pressure gauge.
4. Field measurements are recorded on paper forms or handheld computer.
5. The same procedure is used for measurements referenced to altitude or measuring points, but with a different datum correction.
6. The water level is above land surface but referenced to land-surface datum (LSD). Measurements above LSD are recorded as negative numbers.

Instructions

1. Low-pressure head measurement (direct measurement)
   a. Connect a short length of transparent plastic tubing tightly to the well with hose clamps.
   b. Raise the free end of the tubing until the flow stops.
   c. Rest the measuring scale on the MP.
   d. Place the hose against the measuring scale and read the water level directly. Record the measurement time and WL above MP in the appropriate row of the water-level measurement field form for a low-pressure flowing well measurement (fig. 1)—WL above MP.
   e. Add the MP correction to get the depth to water below LSD. An MP correction above LSD is recorded as a negative number by convention.
   f. Repeat steps b–e for a second check reading. If the check measurement does not agree with the original measurement within 0.1 or 0.2 of a foot, continue to make check measurements until the reason for the lack of agreement is determined or until the results are shown to be repeatable. If more than two readings are taken, use best judgment to select the measurement most representative of field conditions.
2. High-pressure head measurement (indirect measurement)
   a. Make sure that all well valves are closed except the one to the pressure gauge. This will prevent use of the well during the measurement period and assure an accurate water-level reading. Record the original position of each valve that is closed (full open, half open, closed, etc.), so that the well can be restored to its original operating condition.
   b. Connect a flexible hose with a 3-way valve to the well with hose clamps. Expanders/reducers are okay.
   c. Select a gauge where the expected water pressure in the well will fall in the middle third of the gauge range. If in doubt, use a pressure gauge with a 100-pound per square inch (psi) range to make an initial measurement, then select the gauge with the proper range for more accurate measurements.
   d. Attach the pressure gauge to one of the two "open" valve positions using a wrench. Never tighten or loosen the gauge by twisting the case because the strain will disturb the calibration and give erroneous readings.
   e. Bleed air from the hose, using the other "open" valve position.

WATER-LEVEL MEASUREMENT FIELD FORM

Low-Pressure Flowing Well Measurement

SITE INFORMATION

SITE ID (C1)

Equipment ID

Date of Field Visit

Station name (C12)

WATER-LEVEL DATA

	1	2	3	4	5
Time					
WL below MP					
MP correction					
WL below LSD					

Measured by ______________ COMMENTS* ______________

*Comments should include quality concerns and changes in: M.P., ownership, access, locks, dogs, measuring problems, et al.

MEASURING POINT DATA (for MP Changes)

M.P. REMARKS (C324)	BEGINNING DATE (C321)	ENDING DATE (C322)	M.P. HEIGHT (C323) NOTE: (-) for MP below land surface
	month – day – year	– –	.
	– –	– –	.

Final Measurement for GWSI

WATER LEVEL TYPE CODE (C243): L (below land surface), M (below meas. pt.), S (sea level)

DATE WATER LEVEL MEASURED (C235)	TIME (C709)	STATUS (C238)	METHOD (C239)	TYPE (C243)	WATER LEVEL (C237)
month – day – year			M		.

METHOD OF WATER-LEVEL MEASUREMENT(C239)

A	B	C	E	G	H (GWPD12)	L	M (GWPD12)	N	R	S (GWPD1)	T	V (GWPD4)	Z
airline,	analog,	calibrated airline,	estimated,	pressure gage,	calibrated press. gage,	geophysical logs,	manometer,	non-rec. gage,	reported,	steel tape,	electric tape,	calibrated elec. tape	other

SITE STATUS FOR WATER LEVEL (C238)

D	E	F	G	H	I	J	M	N	O	P	R	S	T	V	W	X	Z	BLANK
dry,	recently flowing,	flowing,	nearby flowing	nearby recently flowing,	injector site,	injector site monitor,	plugged,	measurement discon.,	obstruction,	pumping,	recently pumped,	nearby pumping,	nearby recently pumped,	foreign substance,	well destroyed,	surface water effects,	other	static

Figure 1. Water-level measurement field form for low-pressure flowing well measurements. This form, or an equivalent custom-designed form, should be used to record field measurements.

f. Open the pressure gauge valve slowly to reduce the risk of damage by the water-hammer effect to the well, distribution lines, and gauges. Once the needle stops moving, tap the glass face of the gauge lightly with a finger to make sure that the needle is not stuck.

g. Make sure that the well is not being used by checking to see that there are no fluctuations in pressure.

h. Hold the pressure gauge in a vertical position, with the center of the gauge at the exact height of the MP (fig. 2). Read the pressure gauge and record in the Gauge Reading row of the water-level measurement field form for a pressure gauge measurement (fig. 3). Record measurement time.

i. If the pressure gauge has a calibration correction factor, document it in the Gauge Correction row, and record the Corrected Gauge Reading. Multiply by –2.307 under common freshwater temperatures to convert pounds per square inch to feet of water.

j. Apply the MP correction to get the depth to water above LSD. An MP correction above LSD is recorded as a negative number by convention.

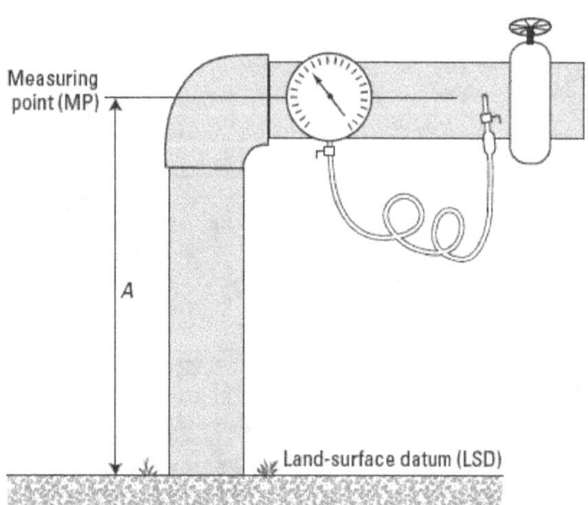

Figure 2. Orientation and position of pressure gauge for measuring water levels in a flowing well.

k. Shut off the well pressure and repeat steps e–i for a second check reading. The measurement should be repeatable within a pressure range based on the range of scale and graduation of the gauge. If more than two readings are taken, use best judgment to select the measurement most representative of field conditions. Document the estimated accuracy of the pressure measurement based on the pressure reading, instrument calibration, the range of the pressure gauge, and manufacturer's guidance.

l. Record the identification number of the pressure gauge with each water-level measurement so that the reading can be back-referenced to the calibration record, if necessary.

Data Recording

All calibration and maintenance data for the pressure gauges are recorded in the calibration logbook. All water-level data are recorded on the water-level measurement field forms (figs. 1 and 2).

References

Cunningham, W.L., and Schalk, C.W., comps., 2011, Groundwater technical procedures of the U.S. Geological Survey, GWPD 3—Establishing a permanent measuring point and other reference marks: U.S. Geological Survey Techniques and Methods 1–A1, 13 p.

Hoopes, B.C., ed., 2004, User's manual for the National Water Information System of the U.S. Geological Survey, Ground-Water Site-Inventory System (version 4.4): U.S. Geological Survey Open-File Report 2005–1251, 274 p.

U.S. Geological Survey, Office of Water Data Coordination, 1977, National handbook of recommended methods for water-data acquisition: Office of Water Data Coordination, Geological Survey, U.S. Department of the Interior, chap. 2, p. 2-11 and 2-12.

WATER-LEVEL MEASUREMENT FIELD FORM
Pressure Gauge Measurement

SITE INFORMATION

SITE ID (C1) | Equipment ID | Date of Field Visit | Station name (C12)

WATER-LEVEL DATA	1	2	3	4	5
Time					
Gauge Reading					
Gauge Correction					
Corrected Gauge Reading					
Conversion to Feet x (−2.307)					
WL below MP					
MP correction					
WL below LSD					

Measured by ______ COMMENTS* ______

*Comments should include quality concerns and changes in: M.P., ownership, access, locks, dogs, measuring problems, et al.

MEASURING POINT DATA (for MP Changes)

M.P. REMARKS (C324) | BEGINNING DATE (C321) month day year | ENDING DATE (C322) | M.P. HEIGHT (C323) NOTE: (-) for MP below land surface

Final Measurement for GWSI

WATER LEVEL TYPE CODE (C243): L below land surface; M below meas. pt.; S sea level

DATE WATER LEVEL MEASURED (C235) month day year | TIME (C709) | STATUS (C238) | METHOD (C239) | TYPE (C243) | WATER LEVEL (C237)

METHOD OF WATER-LEVEL MEASUREMENT(C239)	A	B	C	E	G	H (GWPD12)	L	M	N	R	S (GWPD1)	T	V (GWPD4)	Z
	airline,	analog,	calibrated airline,	estimated,	pressure gage,	calibrated press. gage,	geophysical logs,	manometer,	non-rec. gage,	reported,	steel tape,	electric tape,	calibrated elec. tape	other

SITE STATUS FOR WATER LEVEL (C238)	D	E	F	G	H	I	J	M	N	O	P	R	S	T	V	W	X	Z	BLANK
	dry,	recently flowing,	flowing,	nearby flowing	nearby recently flowing,	injector site,	injector site monitor,	plugged,	measurement discon.,	obstruction,	pumping,	recently pumped,	nearby pumping,	nearby recently pumped,	foreign sub-stance,	well des-troyed,	surface water effects,	other	static

Figure 3. Water-level measurement field form for pressure gauge measurements. This form, or an equivalent custom-designed form, should be used to record field measurements.

GWPD 13—Measuring water levels by use of an air line

VERSION: 2010.1

PURPOSE: To measure the depth to the water surface below a measuring point using the submerged air line method.

Materials and Instruments

1. 1/8 or 1/4-inch diameter, seamless copper tubing, brass tubing, or galvanized pipe with a suitable pipe tee for connecting an altitude or pressure gauge. Flexible plastic tubing also can be used, but is less desirable.
2. Calibrated altitude or pressure gauge, and spare gauges. Gauges that are filled with either oil or silicone work best and are most durable.
3. Compressed air source and corresponding valve stem, usually a Schrader valve. A tire pump can be used on shallow wells and piezometers, but a more substantial source of compressed air is needed where depth to water is hundreds of feet.
4. Small open-end wrench
5. Wire or electrician's tape
6. A steel tape graduated in feet, tenths and hundredths of feet
7. Blue carpenter's chalk
8. Clean rag
9. Field notebook
10. Pencil or pen, blue or black ink. Strikethrough, date and initial errors; no erasures
11. Water-level measurement field form

Data Accuracy and Limitations

1. Accuracy of the water-level measurement is a function of the quality and range of the gauge and the precision to which the length of the air line is known.
2. Water-level measurements using an altitude or pressure gauge can be as accurate as 0.1 foot, but may only be accurate to 1 foot or more, depending on the gauge accuracy and range.
3. Water-level measurements using a pressure gauge are approximate and should not be considered accurate to more than the nearest foot.
4. When measuring deep water levels, corrections for fluid temperatures and vertical differences in air density are additional considerations (Garber and Koopman, 1968).

Advantages

1. Especially useful in pumped wells where water turbulence may preclude using a more precise method.
2. Method can be used while the well is being pumped, when splashing of water makes the wetted-tape method useless.
3. Bends or spirals in the air line do not influence the accuracy of this method as long as the position of the tubing opening is not changed.
4. Can be convenient and is nonintrusive.
5. Air line can be installed once and left in the well for future measurements.

Disadvantages

1. Less accurate than the wetted tape or the electric tape methods.
2. Requires time to install the air line and equipment.
3. Requires careful calculations.

Assumptions

1. An established measuring point (MP) exists and the MP correction length (distance from MP to land-surface datum (LSD)) is known. See GWPD 3 for the technical procedure on establishing a permanent MP.
2. The MP is clearly marked and described so that a person who has not measured the well will be able to recognize it.
3. The air line already is installed, your agency owns the well, or your agency has permission to install the air line.
4. The air line extends far enough below the water level that the lower end remains submerged during pumping of the well.
5. The altitude or pressure gauge and steel tape are calibrated.
6. The same procedure is used for measurements referenced to altitude or measuring points, but with a different datum correction.

Instructions

Figure 1 shows a typical installation for measuring water levels by the air line method.

1. Install an air line pipe or tube in the well. The air line can be installed by either lowering it into the annular space between the pump column and casing after the pump has been installed in the well or by securing it to sections of the pump and pump column with wire or tape as it is lowered into the well.
2. Attach a pipe tee to the top end of the air line. On the opposite end of the pipe tee, attach a Schrader valve stem.
3. Use a wrench to connect an altitude gauge that reads in feet or a pressure gauge that reads in pounds per square inch (psi) to the fitting on top of the pipe tee.
4. Connect a compressed air source to the Schrader valve stem fitting on the pipe tee.
5. Preparatory steps: When pressurizing the air line system (step 8 below), ensure that you supply enough air pressure to purge the water from the air line tubing before a reading is recorded. This can be done by observing the gauge readings while pressurizing the system. After application of pressure, the gauge reading initially will increase to a certain pressure, and when the pressure source is removed, the gauge reading will decrease to a certain pressure. Repeat this process two or three times to ensure that the gauge reads consistently. If the tubing is plugged or crushed, the gauge reading will not decrease after the pressure source is removed. If the tubing is cut or severed, the gauge reading will decrease quickly to zero after the pressure source is removed. In either case, the air line readings will be in error. Also, do not assume that the air line tubing length reported to you is valid. Instead, make water-level measurements by use of steel tape and air line reading simultaneously. This step provides a verified water-level measurement that is relative to the pressure gauge reading. If the two measurements differ, then a correction factor can be calculated. The correction factor will be unique to the well and the gauge.
6. As the water level in the well changes, the gauge reading (h) and the water level below MP (d; fig. 1) must change in a manner such that their sum remains the same. Their sum is a constant (k), which is determined at the same time as a simultaneous wetted-steel tape and pressure gauge measurement is made.
7. To calibrate the air line system, make an initial depth-to-water (d) measurement, with a wetted-steel tape, and an initial air gauge reading (h). Apply any needed correction to the wetted-steel tape measurement. Add d and h to determine the constant value for k. Use the compressed air source to force air into the air line until all the water is expelled from the line. Once all water is displaced from the air line, record the maximum gauge reading.
   - Example 1.—Using an altitude gauge. The initial measured depth to the water level, d, is 25.86 ft; the initial altitude gauge reading, h, is 75.5 ft. Then the constant k = 25.9 ft + 75.5 ft = 101.4 ft (fig. 1).
   - Example 2.—Using a pressure gauge. The initial measured depth to the water level, d, is 85.85 ft; the initial pressure gauge reading, h, is 28 psi. Then the constant k = 86 ft + (2.307 ft/psi x 28 psi) = 86 ft + 64 ft = 150 ft (fig. 1).
8. To measure the water-level depth in a well with an air line, subsequent air line readings are subtracted from the constant k to determine the depth to the water level below the MP. Use a compressed air source to pump compressed air into the air line until all the water is expelled from the line, and record the maximum gauge reading. Apply any correction factor resulting from the calibration process.
   - Example 1.—Depth to the water level in a well using an altitude gauge with a constant k of 101.4 ft. During a later pumping period, the maximum altitude gauge h reads 50.0 ft; therefore, the water level, d, is 101.4 ft – 50.0 ft = 51.4 ft (fig. 2).

- Example 2.—Depth to the water level in a well using a pressure gauge with a constant k of 150 ft. During a later pumping period, the maximum pressure gauge h reads 18 psi; therefore, the water level, d, is 150 ft – (2.307 ft/psi x 18 psi) = 150 ft – 41 ft = 109 ft (fig. 3).

9. Apply the MP correction to get the depth to water below or above LSD.

Data Recording

All data are recorded in the field notebook and on the water-level measurement field forms (fig. 2 or 3) to the appropriate accuracy.

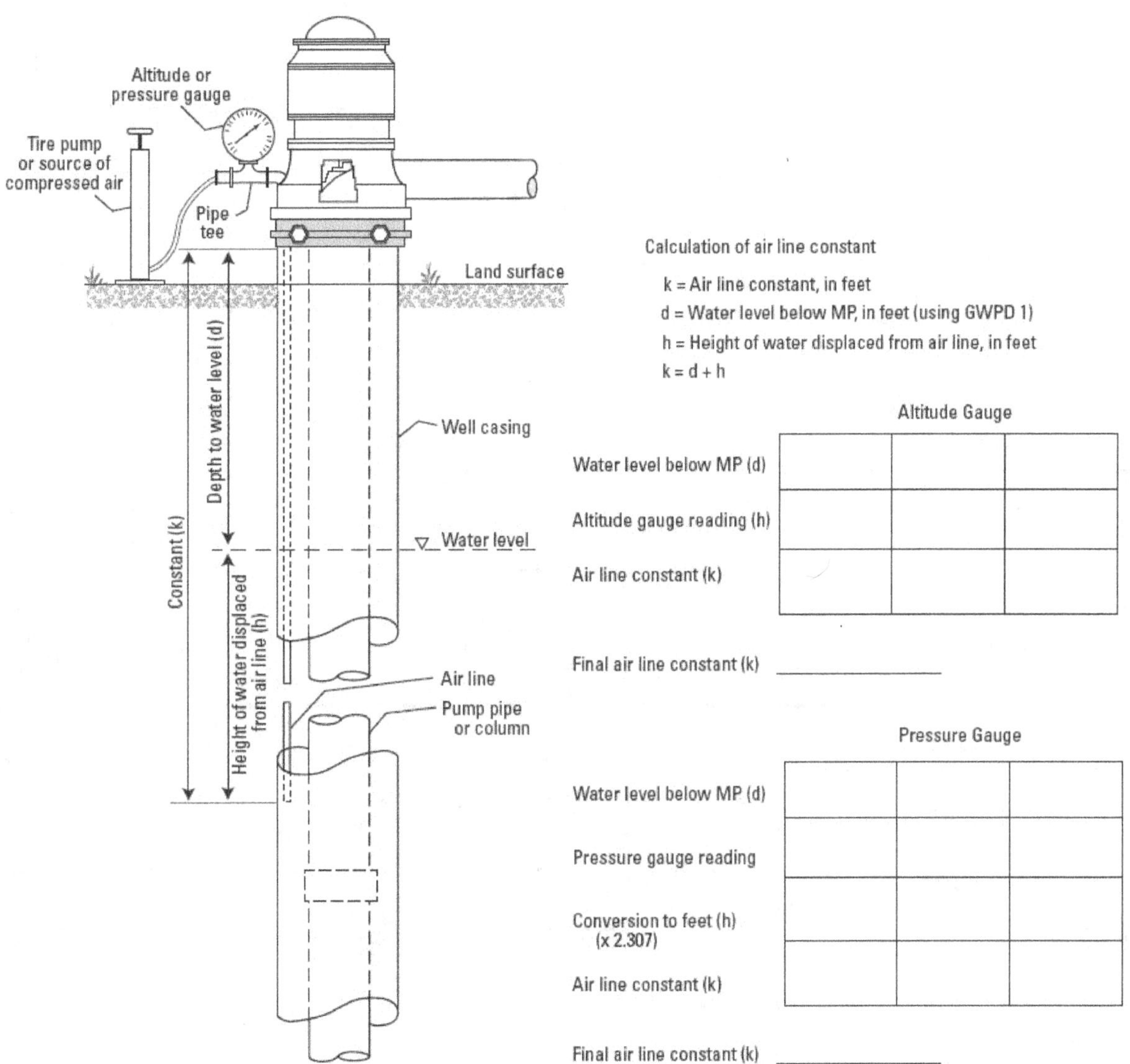

Figure 1. Typical installation for measuring water levels by the air line method and relation of measured depth to water level (d), height of water displaced from air line (h), and constant (k). Constant is calculated by use of altitude gauge or pressure gauge.

WATER-LEVEL MEASUREMENT FIELD FORM
Air Line Measurement: Altitude Gauge

SITE INFORMATION

Equipment ID & Altitude Range ______ Air-line Constant (k) k= ______ Date of Field Visit ______

SITE ID (C1) [| | | | | | | | | | | | | |]

Station name (C12) ______

WATER-LEVEL DATA	1	2	3	4	5
Time					
Gauge Reading					
Gauge Correction					
Corrected Gauge Reading					
Air-Line Constant, K					
WL Below MP					
MP Correction					
WL Above LSD					

Measured by ______ COMMENTS* ______

*Comments should include quality concerns and changes in: M.P., ownership, access, locks, dogs, measuring problems, et al.

MEASURING POINT DATA (for MP Changes)

M.P. REMARKS (C324)	BEGINNING DATE (C321)	ENDING DATE (C322)	M.P. HEIGHT (C323) NOTE: (-) for MP below land surface
______	[ ]-[ ]-[ ] month day year	[ ]-[ ]-[ ]	[ ].[ ]
______	[ ]-[ ]-[ ]	[ ]-[ ]-[ ]	[ ].[ ]

Final Measurement for GWSI

WATER LEVEL TYPE CODE (C243): L M S (L = below land surface, M = below meas. pt., S = sea level)

DATE WATER LEVEL MEASURED (C235)	TIME (C709)	TIME DATUM (C402)	STATUS (C238)	METHOD (C239)	TYPE (C243)	WATER LEVEL (C237)
[ ]-[ ]-[ ] month day year	[ ]	[ ]	[ ]	A	[ ]	[ ].[ ]

GWPD 13

METHOD OF WATER-LEVEL MEASUREMENT (C239)

A	B	C	E	G	H	L	M	N	R	S	T	V	Z
airline,	analog,	calibrated airline,	estimated,	pressure gage,	calibrated press. gage,	geophysical logs,	manometer,	non-rec. gage,	reported,	steel tape,	electric tape,	calibrated elec. tape	other

SITE STATUS FOR WATER LEVEL (C238)

D	E	F	G	H	I	J	M	N	O	P	R	S	T	V	W	X	Z	BLANK
dry,	recently flowing,	flowing,	nearby flowing	nearby recently flowing,	injector site,	injector site monitor,	plugged,	measurement discon.,	obstruction,	pumping,	recently pumped,	nearby pumping,	nearby recently pumped,	foreign substance,	well destroyed,	surface water effects,	other	static

Figure 2. Water-level measurement field form for air line measurement using an altitude gauge. This form, or an equivalent custom-designed form, should be used to record field measurements.

WATER-LEVEL MEASUREMENT FIELD FORM

Air Line Measurement: Pressure Gauge

SITE INFORMATION

Equipment ID & Pressure Range ______________

k = ______ Air-line Constant (k)

Date of Field Visit ______________

SITE ID (C1) | | | | | | | | | | | | | | | |

Station name (C12) ______________

WATER-LEVEL DATA	1	2	3	4	5
Time					
Gauge Reading					
Gauge Correction					
Corrected Gauge Reading					
Conversion to Feet (x 2.307)					
Air-Line Constant, K					
WL Below MP					
MP Correction					
WL Above LSD					

Measured by ______________ COMMENTS* ______________

*Comments should include quality concerns and changes in: M.P., ownership, access, locks, dogs, measuring problems, et al.

MEASURING POINT DATA (for MP Changes)

M.P. REMARKS (C324)	BEGINNING DATE (C321)	ENDING DATE (C322)	M.P. HEIGHT (C323) NOTE: (-) for MP below land surface
______________	__-__-____ (month, day, year)	__-__-____	___.__
______________	__-__-____	__-__-____	___.__

Final Measurement for GWSI

WATER LEVEL TYPE CODE (C243): L (below land surface), M (below meas. pt.), S (sea level)

DATE WATER LEVEL MEASURED (C235)	TIME (C709)	STATUS (C238)	METHOD (C239)	TYPE (C243)	WATER LEVEL (C237)
__-__-____ (month, day, year)	____	_	A	_	___.__

GWPD 13

METHOD OF WATER-LEVEL MEASUREMENT (C239)

A	B	C	E	G	H	L	M	N	R	S	T	V	Z
airline,	analog,	calibrated airline,	estimated,	pressure gage,	calibrated press. gage,	geophysical logs,	manometer,	non-rec. gage,	reported,	steel tape,	electric tape,	calibrated elec. tape	other

SITE STATUS FOR WATER LEVEL (C238)

D	E	F	G	H	I	J	M	N	O	P	R	S	T	V	W	X	Z	BLANK
dry,	recently	flowing,	nearby	nearby	injector	injector	plugged,	measure-	obstruc-	pumping,	recently	nearby	nearby	foreign	well	surface	other	static

Figure 3. Water-level measurement field form for air line measurement using a pressure gauge. This form, or an equivalent custom-designed form, should be used to record field measurements.

References

Cunningham, W.L., and Schalk, C.W., comps., 2011, Groundwater technical procedures of the U.S. Geological Survey, GWPD 3—Establishing a permanent measuring point and other reference marks: U.S. Geological Survey Techniques and Methods 1–A1, 13 p.

Driscoll, F.G., 1986, Groundwater and wells (2d ed.): St. Paul, Minnesota, Johnson Filtration Systems, Inc., 1089 p.

Garber, M.S., and Koopman, F.C., 1968, Methods of measuring water levels in deep wells: U.S. Geological Survey Techniques of Water-Resources Investigations, book 8, chap. A1, p. 6–11.

Hoopes, B.C., ed., 2004, User's manual for the National Water Information System of the U.S. Geological Survey, Ground-Water Site-Inventory System (version 4.4): U.S. Geological Survey Open-File Report 2005–1251, 274 p.

Lohman, S.W., 1953, Measurement of ground-water levels by air-line method: U.S. Geological Survey Open-File Report 53–159, 5 p.

U.S. Geological Survey, Office of Water Data Coordination, 1977, National handbook of recommended methods for water-data acquisition: Office of Water Data Coordination, Geological Survey, U.S. Department of the Interior, chap. 2, p. 2-10.

GWPD 14—Measuring continuous water levels by use of a float-activated recorder

VERSION: 2010.1

PURPOSE: To make continuous water-level measurements in a well using a float-activated recorder.

For some hydrogeologic studies, frequent and uninterrupted water-level measurements may be needed to identify unique properties of the groundwater flow system. In studies in which a more complete picture of water-level fluctuations is needed, automatic float-activated water-level recorders can be installed. Float-activated recorders sense changes in water level by the movement of a weight-balanced float that is lowered into the well.

Materials and Instruments

There are several types of float-activated recording devices. The float or water-level sensing mechanism has not changed much through time. The recording devices have evolved over time from graphical devices to punch tapes to electronic data loggers.

1. Float and non-lead counterweight
2. Small diameter stranded cable or a flat steel tape
3. Graphic recorder, data logger and incremental encoder, integrated data logger/encoder unit, or data collection platform (DCP)
4. Battery, spares, and wiring to connect battery to recording device
5. Tools, including digital multimeter, connectors, crimping tool, and contact-burnishing tool
6. Watch
7. A water-level tape (steel or electric) graduated in hundredths of feet and other materials necessary for depth-to-water measurement
8. Recorder shelter with lock and key
9. Field notebook
10. Pencil or pen, blue or black ink. Strikethrough, date and initial errors; no erasures
11. Water-level measurement field form

Data Accuracy and Limitations

1. The initial water-level setting for a float-activated recorder should be determined using a graduated steel or electric tape which is commonly accurate to 0.01 foot.
2. Each time a float-activated recorder is serviced, calibration check water-level measurements should be made. Data recorded using this procedure are only as accurate as the calibration measurements.
3. Where depth to water is greater than a few feet below the top of the casing, special care should be taken to minimize friction between the float cable and the walls of the well. The float selected should be the largest diameter that can be accommodated by the well casing without excessive friction.
4. Although float-activated recorders can be used successfully in wells that are 2 inches in diameter, in order to avoid friction between the float cable and the walls of the well, 3-inch diameter wells and larger are preferable.
5. Float-activated recorders cannot be used in flowing wells, angled wells, or wells with very deep water levels.

Advantages

1. Graphic recorder

    a. Simplest recording device.

    b. Recorder chart gives a true continuous water-level trace.

    c. Immediate visualization of water-level fluctuations.

    d. Accurate and reliable.

2. Data logger

    a. Stores data in digital form.

    b. Expandable data memory.

    c. Programmable recording intervals.

    d. Accurate and reliable.

3. Data Collection Platform

    a. Provides near real-time data.

    b. Satellite or other transmittal of data.

    c. Accurate and reliable.

    d. Automatic data storage.

Disadvantages

1. Graphic recorder

    a. Limited data-collection time, 1 month versus several months.

    b. Data must be determined manually. Difficult to store in database.

    c. If the graphic recorder clock fails, data will be lost.

    d. This device is archaic, and thus repair is difficult.

2. Data logger and incremental encoders

    a. Rapidly changing water-level peaks may be missed due to programmed preselected time intervals.

    b. Many data loggers require a field computer or a digital interface to download data.

    c. If the memory backup battery fails, data may be lost. Data can be overwritten in some systems.

3. DCP

    a. Transmittal of real-time data can be affected by computer, telephone, or satellite downtime.

    b. Rapidly changing water-level peaks may be missed due to programmed preselected time intervals.

    c. Data transmittal to the satellite can be compromised due to satellite access, tree canopy, ice on antenna, or power supply.

    d. If the memory backup battery fails, data may be lost.

Assumptions

1. A permanent clearly marked measuring point has been established as described in GWPD 3.

2. The user has been trained in making water-level measurements using the graduated steel-tape method as described in GWPD 1, or the electric tape method as described in GWPD 4.

3. Field measurements will be recorded on paper forms. When using a handheld computer to record field measurements, the measurement procedure is the same, but the instructions below refer to specific paper field forms.

Instructions

A wire attached to the float passes over a pulley on the recorder and a counterweight is attached to the other end of the wire and hangs in the well. When the clearance between the float and the well casing is small, the float cable should be set so that the counterweight does not have to pass the float, but is always above or below the water level. If the counterweight is immersed below the water level, a little extra weight should be added to offset the water's buoyancy.

1. The types of float-activated recorders differ by the way in which they record the water level:

    a. Chart or graphic recorder—This type of recorder (fig. 1*A*) is the simplest device, but it is not commonly in use. It is a drum chart that is actuated mechanically by a float that follows the water level. The graphic recorder provides a continuous pen and ink trace of the water level on a chart, which is graduated to record both water level and time. Battery operated clocks for graphic recorders can be set to record a wide variety of intervals, ranging from a few hours to 1 month. The pulley is connected to the recorder drum by gears. A wide range of drum gears are available to set up the chart so that its rotation is proportional to the movement of the float. Figure 1 shows a typical setup for a graphic water-level recorder using a guide pulley assembly (fig. 1*B*) in a small diameter well, as well as a standard position setup (fig. 1*C*). Data are retrieved by changing the paper chart.

b. Data logger and incremental encoder (fig. 2*A*)—Because the data logger and the encoder are separate units connected by a communication cable, this combination of instrumentation allows for a variety of types of equipment to be used. Water-level changes sensed by the float are transferred into a digital signal by the incremental encoder. The digital signal from the incremental encoder is stored on the data logger. This instrumentation suite commonly requires a field computer or a digital interface to download the data.

c. Integrated data logger/incremental encoder units (fig. 2*B*)—This type of recorder combines a data logger and an incremental encoder into one unit. This instrumentation package has replaced the automated digital recorder (ADR punch tape) system. This instrument also requires a field computer or a digital interface to download data.

d. Data collection platform (DCP; fig. 2*C*)—A DCP provides real-time telemeter data using the Geostationary Orbiting Environmental Satellite (GOES) system and can be interfaced with either an incremental encoder or integrated data logger/incremental encoder unit. Data are stored on a data logger and are transmitted to the satellite (GOES) on a fixed schedule (commonly 1 to 4-hour intervals) during a specific time "window." Provided there are no data transmission problems, retrieval of the data is necessary only as a backup. A DCP also may use telephone or other communications technology for data transmission.

2. Select the recording device that best suits the water-level collection needs of the project.

3. Initial installation of the float-activated recorder:

a. Confirm that the well is unobstructed.

b. If the depth of the well is not known, measure the total depth as described in GWPD 11.

c. Install a suitable locking shelter that will protect instruments from weather and vandalism.

d. Establish a measuring point (MP) as described in GWPD 3. Record the MP in the well shelter.

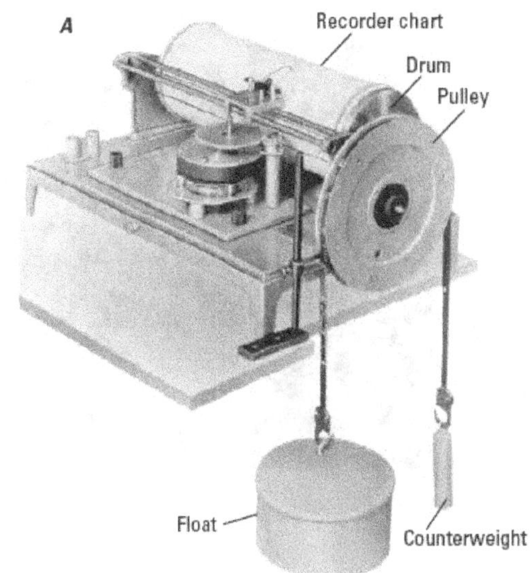

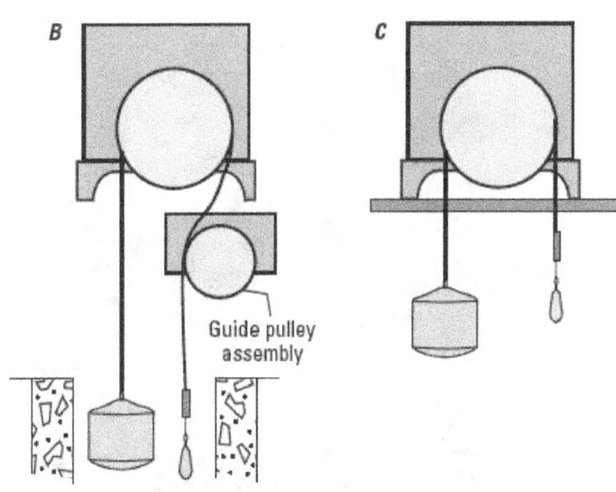

Figure 1. *A*, Standard float-activated graphic water-level recorder (U.S. Bureau of Reclamation, 2001). *B*, Use of a guide pulley assembly to position counterweight inside a small diameter well. *C*, Standard position setup.

e. Measure the depth to water in the well using either GWPD 1 or GWPD 4 to obtain an accurate water-level measurement with which to calibrate the recorder water level (initial calibration). Record the water-level measurement on the Inspection of Continuous Record Well field form (fig. 3).

f. Orient the wheel containing the float tape or float wire and counterweight over the well opening. The float and counterweight must hang freely within the well casing; lack of freedom for the float and counterweight is one of the most common sources of error. The length of float tape or wire should be determined from the expected range of water-level fluctuation; the float should always rest on the water

A. Data logger and incremental encoder

B. Integrated data logger/encoder,

C. Data logger, encoder, and satellite-transmission equipment

Figure 2. *A*, Data logger and incremental encoder. *B*, Integrated data logger/encoder. *C*, Data logger, encoder, and satellite-transmission equipment. Brand names are for illustration purposes only and do not imply endorsement by the U.S. Geological Survey. (Photographs by W.L. Cunningham.)

surface, and the counterweight should always be suspended between the wheel and the water surface. A guide pulley assembly (fig. 1*B*) may be needed for the counterweight. Orient the wheel appropriately, and secure the wheel device and guide pulley assembly to the well shelter to prevent future movement.

g. Balance the float and cable on one side of the pulley against the weight and cable hanging on the opposite side of the pulley. Test the movement of the float wheel by carefully rotating it several inches and releasing it. The tape/recorder should quickly return to the initial value. If it does not return to within 0.01 foot of the initial value, inspect the float tape/wire, float, and counterweight and repair as necessary.

INSPECTION OF CONTINUOUS RECORD WELL
Steel Tape or Calibrated Electric Tape Measurement

SITE INFORMATION

SITE ID (C1) ____________

Measurement Tape ID ________ Date of Field Visit ________

Station name (C12) ________

	1	2	3
Time			
Hold			
Cut			
Tape correction			
WL below MP			
MP correction			
WL below LSD			

Measured by ________

Remarks ________

Barometric Pressure ________ Air Temperature ________

Battery Voltage ________ Replaced? Y / N

Measurement Method: Transducer Float

Checked Float/encoder? Y / N Checked Transducer? Y / N

DATA LOGGER VISIT INFO:

Local time: ________ GMT________ Data logger time: ______

Sensor reading on arrival: ______ Sensor reading on departure: ________ RESET? Y / N

Datum Correction Needed: ________

Retreive data From: ________ To: ________
date/time date/time

Datafile: ________

Remarks: ________

MEASURING POINT DATA (for MP Changes)
M.P. REMARKS (C324)

BEGINNING DATE (C321) [month]-[day]-[year]

ENDING DATE (C322)

M.P. HEIGHT (C323) NOTE: (-) for MP below land surface

Final Measurement for GWSI

DATE WATER LEVEL MEASURED (C235): month - day - year

TIME (C709) STATUS (C238) METHOD (C239) TYPE (C243) WATER LEVEL (C237)

WATER LEVEL TYPE CODE (C243): L M S
L — below land surface; M — below meas. pt.; S — sea level

METHOD OF WATER-LEVEL MEASUREMENT(C239)

A	B	C	E	G	H	L	M	N	R	S (GWPD1)	T	V (GWPD4)	Z
airline,	analog,	calibrated airline,	estimated,	pressure gage,	calibrated press. gage,	geophysical logs,	manometer,	non-rec. gage,	reported,	steel tape,	electric tape,	calibrated elec. tape	other

SITE STATUS FOR WATER LEVEL (C238)

D	E	F	G	H	I	J	M	N	O	P	R	S	T	V	W	X	Z	BLANK
dry,	recently flowing,	flowing,	nearby flowing	nearby recently flowing,	injector site,	injector site monitor,	plugged,	measurement discon.,	obstruction,	pumping,	recently pumped,	nearby pumping,	nearby recently pumped,	foreign substance,	well destroyed,	surface water effects,	other	static

Figure 3. Water-level measurement field form for inspection of continuous record wells. This form, or an equivalent custom-designed form, should be used for continuous recorder inspections and field measurements.

h. Confirm that the direction of the wheel movement is properly recorded (on the display, or by the data logger). For example, when recording depth to water, if the depth to water reading increases as the float is raised, the float was put on in reverse. Correct this error by reversing the direction of the float tape/wire.

i. Set the data logger to the depth to water measured in (e) above using the datum of choice and set the correct time.

j. Measure again to confirm, reset if necessary.

k. Record the water-level measurement on the Inspection of Continuous Record Well field form (fig. 3).

l. Document the equipment serial numbers or other identifiers in the field notebook or on appropriate field forms.

m. Check the battery voltage. Replace if necessary.

n. Confirm that the data logger is operating prior to departure.

4. Subsequent visits to the float-activated recorder:

a. Retrieve groundwater data by using instrument or data logger software.

b. Inspect the equipment to confirm that installation is operating properly. Document the current water level recorded by the sensor (not the most recent water level recorded by the data logger).

c. Measure the depth to water in the well by using either GWPD 1 or GWPD 4 to obtain an accurate water-level measurement with which to check the recorder water level (calibration measurement)

d. Record the water-level measurement on the Inspection of Continuous Record Well field form (fig. 3).

e. Test the movement of the float wheel by carefully rotating it several inches and releasing it. The tape/recorder should return to the same value. If it does not return to within 0.01 foot of the initial value, then inspect the float tape/wire, float, and counterweight and rebalance as necessary

f. Confirm that the direction of the wheel movement is properly recorded (on the display or by the data logger). If the depth to water reading increases as the float is raised, the float was put on in reverse. Correct this error by reversing the direction of the float tape/wire.

g. If the tape measurement differs from the instantaneous instrumentation reading by an amount specified in the groundwater quality assurance procedures of the local office, record it on the inspection sheet and reset the instrumentation to reflect the proper depth to water.

h. Check the battery voltage. Replace if necessary.

i. Make sure the data logger is operating prior to departure.

Data Recording

All data are recorded in the field notebook and on the appropriate field form.

References

Bureau of Reclamation, 2001, Water measurement manual, A water resources technical publication (2d ed. rev. reprinted): U.S. Department of the Interior, 485 p., accessed December 17, 2010, at *http://www.usbr.gov/pmts/hydraulics_lab/pubs/wmm/*.

Garber, M.S., and Koopman, F.C., 1968, Methods of measuring water levels in deep wells: U.S. Geological Survey Techniques of Water-Resources Investigations, book 8, chap. A1, 23 p.

Cunningham, W.L., and Schalk, C.W., comps., 2011a, Groundwater technical procedures of the U.S. Geological Survey, GWPD 1—Measuring water levels by use of a graduated steel tape: U.S. Geological Survey Techniques and Methods 1–A1, 4 p.

Cunningham, W.L., and Schalk, C.W., comps., 2011b, Groundwater technical procedures of the U.S. Geological Survey, GWPD 3—Establishing a permanent measuring point and other reference marks: U.S. Geological Survey Techniques and Methods 1–A1, 13 p.

Cunningham, W.L., and Schalk, C.W., comps., 2011c, Ground-Water technical procedures of the U.S. Geological Survey, GWPD 4—Measuring water levels by use of an electric tape: U.S. Geological Survey Techniques and Methods 1–A1, 6 p.

U.S. Geological Survey, Office of Water Data Coordination, 1977, National handbook of recommended methods for water-data acquisition: Office of Water Data Coordination, Geological Survey, U.S. Department of the Interior, chap. 2, p. 2-12–2-14.

GWPD 15—Obtaining permission to install, maintain, or use a well on private property

VERSION: 2010.1

PURPOSE: To describe a procedure for properly obtaining permission to install, maintain, or use a well on private property, for activities such as geophysical explorations, water-level monitoring, and collection of water samples.

U.S. Geological Survey (USGS) policy for access to private lands is governed by Chapter 500.11 in the Survey Manual. It is USGS policy to obtain written permission before drilling, collecting groundwater samples, maintaining a continuous recorder, or making a groundwater-level measurement on private property, restricted public property, and leased Federal land. Test drilling and data collection preferably should be confined to public lands (Federal, State, county, or municipally owned) when the location will serve as well as one on privately owned land. However, if the information needed can be obtained only at a site on private property, that site may be used if permission to drill test wells, sample, or operate observation wells is obtained in advance.

Materials and Instruments

1. Form 9-1483, Well Drilling/Sampling Agreement
2. Permission to Collect Water Samples form
3. Form 9-3106, Well Transfer Agreement
4. Site location map
5. Field notebook
6. Pencil or pen, blue or black ink. Strikethrough, date, and initial errors; no erasures

Data Accuracy and Limitations

When public land is not suitable, the use of private property is permitted if, prior to drilling, sampling, or data collection operations, a signed agreement for access to and installation, maintenance, and use of the test hole or observation well is obtained from the property owner.

Assumptions

1. Needed information can be collected only at a site on private property.
2. The person requesting permission to install, maintain, or use a well on private property is familiar with Office of Ground Water Technical Memorandum 2003.03 and associated policies.
3. The requestor is also familiar with State law requirements to notify the local One Call Center (in some States referred to as, "call before you dig") before blasting, boring, digging, drilling, trenching, or other earth moving operations.

Instructions

1. If seeking permission to drill: Complete all the information on the Well Drilling/Sampling Agreement form (fig. 1, Form 9-1483). Attach to the agreement a site map showing the location of each proposed test hole and (or) observation well. Form 9-1483 must be signed by the landowner and a USGS representative.
    a. Each agreement is assigned a number consisting of the first four digits of the cost center, hyphen, a sequential number beginning with 01, and the year in which the agreement is processed. For example, 4563-0110.
    b. Form 9-1483 or an equivalent form must be signed by the landowner and a USGS representative.
    c. When work at a well is completed and the conditions outlined in Office of Ground Water Technical Memorandum 2003.03 are met, ownership of a well may

Tips on Help Using This Form
Form 9-1483
Revised (October 2002)

Agreement Number:

Well Drilling/Sampling Agreement

The landowner agrees that the U.S. Geological Survey (USGS), ______ District may install and maintain a monitoring well on the landowner's property at a mutually agreed-upon site at the location listed below. The landowner also agrees that the USGS will have access to the site, as it reasonably deems necessary for water-level measurements, geophysical measurements and/or water-quality sampling purposes during the life of this agreement.

The monitoring well will be a hole extending into the earth produced by drilling or auguring. The hole may be cased and screened at an appropriate depth for water level measurements and/or sampling. The well water may be sampled for multiple constituents.

Excavation and/or installation of the well may begin at any time after this agreement is fully executed. The well shall be excavated, installed, and properly maintained by the USGS at its own expense. This agreement shall be regarded as granting a license or easement, whichever may be most appropriately characterizes it under applicable state law, in favor of USGS to enter landowner's property for the purposes noted herein.

At the expiration of this Agreement, the well may be abandoned in one of the following ways:

1. The well may be removed, filled and/or plugged, according to federal, state, and local regulations, by the USGS at its own expense within a reasonable time after the expiration of this Agreement. The USGS, soon thereafter, shall restore the property, again at its own expense, as nearly as possible to the same condition as existed prior to the excavation and/or installation of the well, or

2. At the request of the Landowner, and if the well has been in existence for five years or more, ownership of the well may be transferred to the Landowner under a separate Well Transfer Agreement.

During the life of this Agreement, the Federal Government will be liable for any loss related to the installation, operation, maintenance, or other activities associated with the well described above in accordance with, and to the extent permitted, under the Federal tort Claims Act (28 U.S.C. 1346(b) and 2671 *et seq.*).

This agreement shall be come effective when fully executed and shall continue in force for 5 years unless terminated earlier by the USGS upon 60 days written notice. After 5 years, the Agreement remains in force until terminated by either the USGS or the Landowner upon 60 days written notice to the other party.

Description of well located at Lat. ______ Long. ______ (Attach Drawing)

Landowner:

Address:

Tel. Number: ______ Fax Number ______

USGS Center Director:

Address:

Tel. Number: ______ Fax Number ______

USGS Project Chief

Tel. Number: ______ Fax Number ______

U.S. GEOLOGICAL SURVEY
By:

______ Date: ______

(Name)

LANDOWNER
By:

______ Date: ______

Notary Seal:

Figure 1. Well Drilling/Sampling Agreement, Form 9-1483.

As consideration for the rights and privileges granted herein, the USGS shall make a one-time payment to the Landowner in the sum of $ ______. This Agreement shall be binding upon Landowner's devises, heirs, successors, and assigns.

Landowner:

Address:

Tel. Number: Fax Number:

USGS Center Director:

Address:

Tel. Number: Fax Number:

USGS Project Chief:

Tol. Number: Fax Number:

U.S. GEOLOGICAL SURVEY
By

______________ Date: ______________

(Name)

LANDOWNER
By

______________ Date: ______________

Notary Seal:

Figure 1. Well Drilling/Sampling Agreement, Form 9-1483.—Continued

be transferred back to the landowner. Form 9-3106 (fig. 2, Well Transfer Agreement) provides for transfer of well ownership. Form 9-3106 must be signed by the landowner and a USGS representative.

2. If seeking permission to collect water samples from a well: Experience has shown that oral permission to collect water samples is easier to obtain, but written permission provides stronger legal protection. Form 9-1483 includes permission for the USGS to take water-quality samples from a well being drilled. However, if an existing well is used instead of drilling a well, use of the Permission to Collect Water Samples form (fig. 3) is warranted. Strong consideration should be used to incorporate this form even when Form 9-1483 is in place. Figure 2 or an equivalent form must be signed by the permitter (landowner) and a USGS representative.

3. If seeking permission to maintain a continuous recorder or make a groundwater-level measurement on private property, restricted public property, or leased Federal land: **The USGS preferred business practice is that permission for this activity be obtained in writing using Form 9-1483 or equivalent. Long-standing oral agreements and oral agreements made in situations where obtaining written permission would be prohibitive can be documented by using the form** shown in figure 4 (Format for Letter Requesting Permission To Enter Private Property) or by obtaining the information included in figure 5 (Documentation of Oral Permission to Access Private Lands) and documenting the oral permission as soon as possible.

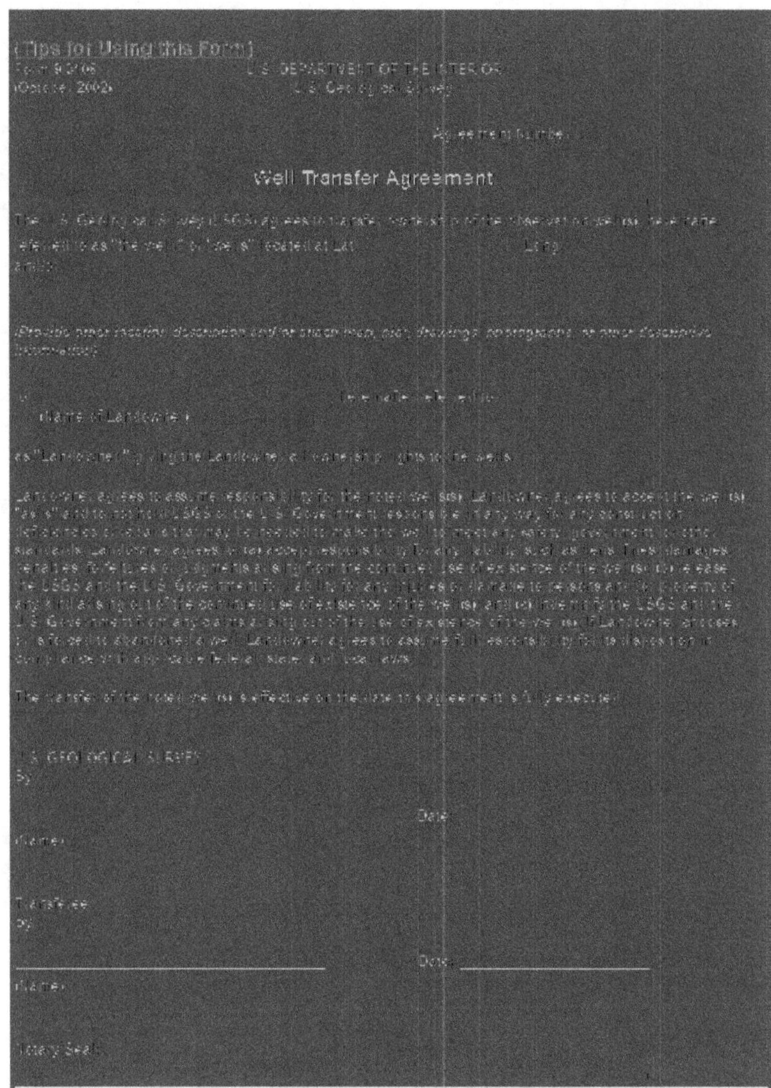

(Tips for Using this Form)

U.S. DEPARTMENT OF THE INTERIOR

Well Transfer Agreement

Date

Date

Figure 2. Well Transfer Agreement Form 9-3106 for transfer of well ownership.

Unnumbered form (from WRD Memo No. 90.34)

U.S. GEOLOGICAL SURVEY

Permission to Collect Water Samples.

I (we) ____ hereby give my (our) permission to the U.S. Geological Survey to collect a water sample (s) from my well, spring, stream, lake, or reservoir. I understand that this sample will be analyzed by the U.S. Geological Survey and that the data will be used for scientific purposes. I also understand that I will be furnished a copy of the analysis and that the data will be stored in the Geological Survey's computer storage files and become public information at that time. The U.S. Geological Survey has also informed me (us) that some results of the analysis that exceed the U.S. Environmental Protection Agency's Primary Drinking Water Standard Maximum Contaminant Levels may be reported to a local, State, or Federal regulatory agency.

In addition to collecting a sample (s) for a laboratory analysis, the U.S. Geological Survey may also make a series of concurrent physical measurements such as water level, streamflow, pH, and temperature.

If I (we) have any questions about this program of the U.S. Geological Survey,

I can contact ____

At the following telephone number ________________________________

____

Signature, Permitter Date

____

Signature, U.S. Geological Survey Date

Local address

____

____

____

Figure 3. Form to use to obtain permission to collect water samples.

U.S. Geological Survey Manual

Figure 500.11.1

Format for Letter Requesting Permission To Enter Private Property (to be printed on Official Letterhead)

(Insert Name of Private Landowner)
(Insert Address of Private Landowner)

(Insert Date)

Dear (Insert Name of Private Landowner):

The U.S. Geological Survey requires employees to obtain written permission from landowners in certain cases before entering onto private property to conduct new surveys or scientific sampling. Consequently, we are hereby requesting your approval to enter your land for the purpose described below. The data and/or samples collected will be used for scientific purposes and will be provided to you upon request.

Specific information regarding this request is as follows:

1. (proposed date and time of entry and departure, or period of time during which recurring visits will be necessary).

2. (kind and number of vehicles to be used).

3. (number of persons in the party).

4. (name, office address, and contact information of chief of party).

5. (purpose of the work).

6. (locations on the property where work is to be done).

7. (approximate frequency of aircraft flights along lines of sight for temperature and pressure measurements, in connection with geodimeter or similar work, if applicable).

We will make every effort to minimize disturbance or disruption to your property. However, in the unlikely event that property damage results, you are entitled to file a claim to recover your damages (tort claim). Please contact (insert name and telephone number of tort claims contact) immediately if property damage should occur.

If you have any questions about this program of the U.S. Geological Survey, you may contact (insert name of chief of project) at the following telephone number: (insert number).

If you consent to this request, please sign below and (list method of return, e.g., envelope provided, leave at a designated location, etc.). Thank you for your cooperation.

Sincerely,

____

(Signature and Printed Name of Requestor)

Approval: ____

Landowner Signature Date

Figure 4. Format for letter requesting permission to enter private property (U.S. Geological Survey Manual 500.11).

U.S. Geological Survey Manual

Figure 500.11.2
Documentation of Oral Permission to Access Private Lands

The U.S. Geological Survey obtained oral permission to access private lands as follows:

Description of the work and/or project title, to include date and time of entry and departure or anticipated duration of the work if recurring visits will be made:

Printed name and address of landowner contacted:

____ The landowner was provided with the following information:

1. (proposed date and time of entry and departure, or period of time during which recurring visits will be necessary).

2. (kind and number of vehicles to be used).

3. (number of persons in the party).

4. (name, office address, and contact information of chief of party).

5. (purpose of the work).

6. (locations on the property where work is to be done).

7. (approximate frequency of aircraft flights along lines of sight for temperature and pressure measurements, in connection with geodimeter or similar work, if applicable).

Date permission was granted:

Office location of initiating party:

Name and signature of member of field party who obtained permission:

Other persons in the party who witnessed the oral permission (as applicable):

The documentation of an oral agreement should be retained in the project file by the initiating office until the project is completed and in accordance with the *Handbook for Managing USGS Records*, 432-1-H.

Figure 5. Documentation of oral permission to access private lands (U.S. Geological Survey Manual 500.11).

Data Recording

Permission details are recorded on the associated forms. The original form is kept in the office, and a copy is included in the well folder that is brought to the field.

The Agency Use Code (C803) on the Groundwater Site Schedule (Form 9-1904-A; fig. 6) should be used to indicate the type of agreement in place. If the well is not owned by the USGS, use codes A, L, or M when coding the site in the National Water Information System. For further information, refer to USGS Water Resources Discipline Policy Memorandum 2009.02.

Agency use code (C803)	Short description	Long description
A	Active - no/na	Active data collection site with undocumented or unneeded landowner agreement
L	Active - Written	Active data collection site with written landowner agreement (Form 9-1483)
M	Active - Oral	Active data collection site with memo documenting oral landowner agreement

Reference

U.S. Geological Survey, 2003, Agreement forms for gaging station and observation well installations and transfers: Office of Ground Water Technical Memorandum 2003.03, accessed December 17, 2010, at *http://water.usgs.gov/admin/memo/GW/gw03.03.html.*

U.S. Geological Survey, 2008, U.S. Geological Survey Manual 500.11—Obtaining permission for access to private lands, accessed December 17, 2010, at *http://www.usgs.gov/usgs-manual/500/500-11.html.*

U.S. Geological Survey, 2009, Maintaining an auditable record of USGS discontinued water monitoring station liabilities: Water Resources Discipline Policy Memorandum No. 2009.02, accessed at *http://water.usgs.gov/admin/memo/policy/wrdpolicy09.02.pdf.*

FORM NO. 9-1904-A
Revised Sept 2009, NWIS 4.9

File Code ______
Date ______

Coded by ______
Checked by ______
Entered by ______

U.S DEPT. OF THE INTERIOR
GEOLOGICAL SURVEY

GROUNDWATER SITE SCHEDULE
General Site Data

AGENCY CODE (C4) USGS | SITE ID (C1) | PROJECT (C5)

STATION NAME (C12/900)

[1] SITE TYPE (C802) Primary - Secondary | DISTRICT (C6) | COUNTRY (C41) | STATE (C7)

COUNTY or TOWN (C8) ______ County code

LATITUDE (C9) | LONGITUDE (C10) | LAT/LONG ACCURACY (C11): H Hndrth sec., 1 tenth sec., 5 half sec., S sec., R 3 sec., F 5 sec., T 10 sec., M min., U Unknown

LAT/LONG METHOD (C35): C land net, D DGPS, G GPS, L LORAN, M map, N interpolated digital map, R reported, S survey, U unknown

LAT/LONG DATUM (C36): NAD27 North American Datum of 1927, NAD83 North American Datum of 1983

ALTITUDE (C16)

ALTITUDE ACCURACY (C18)

ALTITUDE METHOD (C17): A altimeter, D DGPS, G GPS, I IfSAR, J LIDAR, L Level, M map, N DEM, R reported, U unknown

ALTITUDE DATUM (C22): NGVD29 National Geodetic Vertical Datum of 1929, NAVD88 North American Vertical Datum of 1988

LAND NET (C13): ¼ ¼ ¼ S section, T township, range, merid

TOPOGRAPHIC SETTING (C19): A alluvial fan, B playa, C stream channel, D depression, E dunes, F flat, G flood-plain, H hill-top, K sink-hole, L lake or swamp, M mangrove swamp, O off-shore, P pediment, S hill-side, T terrace, U undulating, V valley flat, W upland draw

HYDROLOGIC UNIT CODE (C20) | DRAINAGE BASIN CODE (C801) | STANDARD TIME ZONE (C813) | DAYLIGHT SAVINGS TIME FLAG (C814) Y OR N

MAP NAME (C14) | MAP SCALE (C15)

AGENCY USE (C803): A active no/na, D discontinued, I inactive site, L active written, M active oral, O inventory site, R remediated

[2] NATIONAL WATER-USE (C39)

DATA TYPE (C804) Place an 'A' (active), an 'I' (inactive), or an 'O' (inventory) in the appropriate box: WL cont, WL int, QW cont, QW int, PR cont, PR int, EV cont, EV int, wind vel., tide cont, tide int, sed. con, sed. ps, peak flow, low flow, state water use

INSTRUMENTS (C805) (Place a "Y" in the appropriate box): digital recorder, graphic recorder, telemetry land line, telemetry radio, telemetry satellite, AHDAS, crest-stage gage, tide gage, deflection meter, bubble gage, stilling well, CR type recorder, weighing rain gage, tipping bucket rain gage, acoustic velocity meter, electromagnetic flowmeter, pressure transducer

DATE INVENTORIED (C711) month - day - year

RECORD READY FOR WEB (C32): Y ready to display, C conditional, P proprietary, L local use only

REMARKS (C806)

FOOTNOTES

1 SITE TYPE (C802)

GL	Glacier	OC	Ocean	GW	Well	SB	Subsurface
WE	Wetland	OC-CO	Coastal	GW-CR	Collector or Ranney type well	SB-CV	Cave
AT	Atmosphere	LK	Lake, Reservoir, Impoundment	GW-EX	Extensometer well	SB-GWD	Groundwater drain
ES	Estuary			GW-HZ	Hyporheic -zone well	SB-TSM	Tunnel, shaft, or mine
LA	Land	SP	Spring	GW-IW	Interconnected wells	SB-UZ	Unsaturated zone
LA-EX	Excavation	ST	Stream	GW-TH	Test hole not completed as a well		
LA-OU	Outcrop	ST-CA	Canal	GW-MW	Multiple wells		
LA-SNK	Sinkhole	ST-DCH	Ditch				
LA-SH	Soil hole	ST-TS	Tidal stream				
LA-SR	Shore	FA-WIW	Waste-Injection well				

2 WS water supply, DO domestic, CO commercial, IN industrial, IR irrigation, MI mining, LV livestock, PH power hydro-electric, ST waste water treatment, RM remediation, TE thermo-electric power, AQ aquaculture

C22 Other (see manual for codes)
C36 Other (see manual for codes)
C39 is mandatory for all sites having data in SWUDS.

Figure 6. Groundwater Site Schedule, Form 9-1904-A.

GENERAL SITE DATA

DATA RELIABILITY (C3): C (field checked), L (poor location), M (minimal data), U (unchecked)

DATE OF FIRST CONSTRUCTION (C21): month – day – year

USE OF SITE (C23): A (anode), C (standby emer. supply), D (drain), E (geothermal), G (seismic), H (heat reservoir), M (mine), O (observation), P (oil or gas), R (recharge), S (repressurize), T (test), U (unused), V (withdrawal/return), W (withdrawal), X (waste), Z (destroyed)

SECONDARY USE OF SITE (C301) (See use of site)

TERTIARY USE OF SITE (C302) (See use of site)

USE OF WATER (C24): A (air cond.), B (bottling), C (commercial), D (dewater), E (power), F (fire), H (domestic), I (irrigation), J (industrial (cooling)), K (mining), M (medicinal), N (industrial), P (public supply), Q (aquaculture), R (recreations), S (stock), T (institutional), U (unused), Y (desalination), Z (other)

SECONDARY USE OF WATER (C25) (see use of water)

TERTIARY USE OF WATER (C26) (see use of water)

AQUIFER TYPE (C713): U (unconfined single), N (unconfined multiple), C (confined single), M (confined multiple), X (mixed)

PRIMARY AQUIFER (C714)

NATIONAL AQUIFER (C715)

HOLE DEPTH (C27)

WELL DEPTH (C28)

SOURCE OF DEPTH DATA (C29): A (other gov't), D (driller), G (geologist), L (logs), M (memory), O (owner), R (other reported), S (reporting agency), Z (other)

WATER-LEVEL DATA

DATE WATER-LEVEL MEASURED (C235): month – day – year

TIME (C709)

WATER-LEVEL TYPE CODE (C243): L (land surface), M (meas. pt.), S (vertical datum)

WATER LEVEL (C237/241/242)

MP SEQUENCE NO. (C248) (Mandatory if WL type=M)

WATER-LEVEL DATUM (C245) (Mandatory if WL type=S): NGVD29 (National Geodetic Vertical Datum Of 1929), NAVD88 (North American Vertical Datum Of 1988), Other (See manual for codes)

SITE STATUS FOR WATER LEVEL (C238): A (atmos. pressure), B (tide stage), C (ice), D (dry), E (recently flowing), F (flowing), G (nearby flowing), H (nearby recently flowing), I (injector site), J (injector site monitor), M (plugged), N (measurement discontinued), O (obstruction), P (pumping), R (recently pumped), S (nearby pumping), T (nearby recently pumped), V (foreign substance), W (well destroyed), X (affected by surface water), Z (other)

METHOD OF WATER-LEVEL MEASUREMENT (C239): A (airline), B (analog), C (calibrated airline), D (differential GPS), E (estimated), F (transducer), G (pressure gage), H (calibrated press. gage), L (geophysical logs), M (manometer), N (non-rec. gage), O (observed), P (acoustic pulse), R (reported), S (steel tape), T (electric tape), V (calibrated elec. tape), Z (other)

WATER-LEVEL ACCURACY (C276): 0 (foot), 1 (tenth), 2 (hundredth), 9 (not to nearest foot)

SOURCE OF WATER-LEVEL DATA (C244): A (other gov't), D (driller's log), G (geologist), L (geophysical logs), M (memory), O (owner), R (other reported), S (reporting agency), Z (other)

PERSON MAKING MEASUREMENT (C246) (WATER LEVEL PARTY)

MEASURING AGENCY (C247) (SOURCE)

EQUIP ID (C249) (20 char)

REMARKS (C267) (256 char)

RECORD READY FOR WEB (C858): Y (ready to display), C (conditional), P (proprietary), L (local use only)

CONSTRUCTION DATA

RECORD TYPE (C754): CONS

RECORD SEQUENCE NO. (C723)

DATE OF COMPLETED CONSTRUCTION (C60): month – day – year

NAME OF CONTRACTOR (C63)

SOURCE OF DATA (C64): A (other gov't), D (driller), G (geologist), L (logs), M (memory), O (owner), R (other reported), S (reporting agency), Z (other)

METHOD OF CONSTRUCTION (C65): A (air-rotary), B (bored or augered), C (cable tool), D (dug), H (hydraulic rotary), J (jetted), P (air percussion), R (reverse rotary), S (sonic), T (trenching), V (driven), W (drive wash), Z (other)

TYPE OF FINISH (C66): C (porous concrete), F (gravel w/perf.), G (gravel screen), H (horiz. gallery), O (open end), P (perf or slotted), S (screen), T (sand point), W (walled), X (open hole), Z (other)

TYPE OF SEAL (C67): B (bentonite), C (clay), G (cement grout), N (none), Z (other)

BOTTOM OF SEAL (C68)

METHOD OF DEVELOPMENT (C69): A (air-lift pump), B (bailed), C (compressed air), J (jetted), N (none), P (pumped), S (surged), Z (other)

HOURS OF DEVELOPMENT (C70)

SPECIAL TREATMENT (C71): C (chemicals), D (dry ice), E (explosives), F (deflocculent), H (hydrofracturing), M (mechanical), Z (other)

2 - Groundwater Site Schedule

CONSTRUCTION HOLE DATA (3 sets shown)

RECORD TYPE (C756) HOLE

RECORD SEQUENCE NO. (C724)

SEQUENCE NO. OF PARENT RECORD (C59)

DEPTH TO TOP OF INTERVAL (C73)

DEPTH TO BOTTOM OF INTERVAL (C74)

DIAMETER OF NTERVAL (C75)

RECORD SEQUENCE NO. (C724)

DEPTH TO TOP OF INTERVAL (C73)

DEPTH TO BOTTOM OF INTERVAL (C74)

DIAMETER OF NTERVAL (C75)

RECORD SEQUENCE NO. (C724)

DEPTH TO TOP OF INTERVAL (C73)

DEPTH TO BOTTOM OF INTERVAL (C74)

DIAMETER OF NTERVAL (C75)

CONSTRUCTION CASING DATA (4 sets shown)

RECORD TYPE (C758) CSNG

RECORD SEQUENCE NO. (C725)

SEQUENCE NO. OF PARENT RECORD (C59)

DEPTH TO TOP OF CASING (C77)

DEPTH TO BOTTOM OF CASING (C78)

DIAMETER OF CASING (C79)

4 CASING MATERIAL (C80)

CASING THICKNESS (C81)

RECORD SEQUENCE NO. (C725)

SEQUENCE NO. OF PARENT RECORD (C59)

DEPTH TO TOP OF CAS NG (C77)

DEPTH TO BOTTOM OF CAS NG (C78)

DIAMETER OF CASING (C79)

4 CAS NG MATERIAL (C80)

CASING THICKNESS (C81)

RECORD SEQUENCE NO. (C725)

SEQUENCE NO. OF PARENT RECORD (C59)

DEPTH TO TOP OF CASING (C77)

DEPTH TO BOTTOM OF CASING (C78)

DIAMETER OF CASING (C79)

4 CASING MATERIAL (C80)

CASING THICKNESS (C81)

RECORD SEQUENCE NO. (C725)

SEQUENCE NO. OF PARENT RECORD (C59)

DEPTH TO TOP OF CASING (C77)

DEPTH TO BOTTOM OF CASING (C78)

DIAMETER OF CASING (C79)

4 CASING MATERIAL (C80)

CASING THICKNESS (C81)

FOOTNOTE:

4 CASING MATERIAL CODES

A	B	C	D	E	F	G	H	I	J	K	L	M	N	P	Q	R	S	T	U	V	W	X	Y	Z	4	6
abs	brick	concrete	copper	PTFE	Fiber-glass	galv. iron	Fiber-glass plastic	wrought iron	Fiber-glass epoxy	PVC thread-ed	glass	other metal	PVC glued	PVC or plastic	FEP	rock or stone	steel	tile	coated steel	stain-less steel	wood	steel carbon	steel galva-nized	other mat.	stain-less 304	stain-less 316

CONSTRUCTION OPENINGS DATA (3 sets shown)

RECORD TYPE (C760) O P E N   RECORD SEQUENCE NO. (C726)   SEQUENCE NO. OF PARENT RECORD (C59)

DEPTH TO TOP OF INTERVAL (C83) .   DEPTH TO BOTTOM OF INTERVAL (C84) .   DIAMETER OF INTERVAL (C87) .

[5] MATERIAL TYPE (C86)   [6] TYPE OF OPENING (C85)   LENGTH OF OPENING (C89) .   WIDTH OF OPENING (C88) .

RECORD SEQUENCE NO. (C726)

DEPTH TO TOP OF INTERVAL (C83) .   DEPTH TO BOTTOM OF INTERVAL (C84) .   DIAMETER OF INTERVAL (C87) .

[5] MATERIAL TYPE (C86)   [6] TYPE OF OPENING (C85)   LENGTH OF OPENING (C89) .   WIDTH OF OPENING (C88) .

RECORD SEQUENCE NO. (C726)

DEPTH TO TOP OF INTERVAL (C83) .   DEPTH TO BOTTOM OF INTERVAL (C84) .   DIAMETER OF INTERVAL (C87) .

[5] MATERIAL TYPE (C86)   [6] TYPE OF OPENING (C85)   LENGTH OF OPENING (C89) .   WIDTH OF OPENING (C88) .

FOOTNOTES:

[5] TYPE OF MATERIAL CODES FOR OPEN SECTIONS

A	B	C	D	E	F	G	H	I	J	K	L	M	N	P	Q	R	S	T	V	W	X	Y	Z	4	6
ABS	brass or bronze	concrete	ceramic	PTFE	fiber-glass	galv. iron	fiber-glass plastic	wrought iron	fiber-glass epoxy	PVC thread-ed	glass	other metal	PVC glued	PVC	FEP	stain-less steel	steel	tile	brick	mem-brane	steel carbon	steel galva-nized	other	stain-less 304	stain-less 316

[6] TYPE OF OPENINGS CODES

F	L	M	P	R	S	T	W	X	Z
fractured rock	louvered or shutter-type	mesh screen	perforated, porous or slotted	wire-wound screen	screen (unk.)	sand point screen	walled or shored	open hole	other

CONSTRUCTION MEASURING POINT DATA

RECORD TYPE (C766) M P N T   RECORD SEQUENCE NO. (C728)   BEGINNING DATE (C321) — — (month, day, year)   ENDING DATE (C322) — —

M.P. HEIGHT (C323) .   ALTITUDE OF MEASURING POINT (C325)   ALTITUDE METHOD (C326)   ALTITUDE ACCURACY (C327)

ALTITUDE DATUM (C328)   M.P. REMARKS (C324)

RECORD READY FOR WEB (C857)

Y	C	P	L
ready to display	condi-tional	proprie-tary	local use only

CONSTRUCTION LIFT DATA

RECORD TYPE (C752) L I F T

RECORD SEQUENCE NO. (C254)

TYPE OF LIFT (C43)

A	B	C	J	P	R	S	T	U	X	Z
air	bucket	centri-fugal	jet	piston	rotary	submer-sible	turbine	un-known	no lift	other

DATE RECORDED (C38) month – day – year

PUMP INTAKE DEPTH (C44)

TYPE OF POWER (C45)

D	E	G	H	L	N	S	W	Z
diesel	electric	gaso-line	hand	LP gas	natural gas	solar	windmill	other

HORSE-POWER RATING (C46) .

MANUFACTURER (C48)

SERIAL NO. (C49)

POWER COMPANY (C50)

POWER COMPANY ACCOUNT NUMBER (C51)

POWER METER NUMBER (C52)

PUMP RATING (C53) (million gallons/units of fuel) .

ADDITIONAL LIFT (C255)

PERSON OR COMPANY MAINTAINING PUMP (C54)

RATED PUMP CAPACITY (gpm) (C268)

STANDBY POWER (C56) (see TYPE OF POWER)

HORSEPOWER OF STANDBY POWER SOURCE (C57) .

MISCELLANEOUS OWNER DATA

RECORD TYPE (C768) O W N R

RECORD SEQUENCE NO. (C718)

DATE OF OWNERSHIP (C159) – –

WU OWNER TYPE (C350)

CP	GV	IN	MI	OT	TG	WS
Corporation	Govern-ment	Individual	Military	Other	Tribal	Water Supplier

END DATE OF OWNERSHIP (C374) – –

OWNER'S NAME (C161)

EXAMPLES: JONES, RALPH A.
JONES CONSTRUCTION COMPANY

OWNER'S PHONE NUMBER (C351)

ACCESS TO OWNER'S NAME (C352)

0	1	2	3	4
Public Access	Coop-erator	USGS Only	District Only	Proprietary

OWNER'S ADDRESS (LINE 1) (C353)

OWNER'S ADDRESS (LINE 2) (C354)

OWNER'S CITY NAME (C355)

STATE (C356)

OWNER'S ZIP CODE (C357) –

OWNER'S COUNTRY NAME (C358)

ACCESS TO OWNER'S PHONE/ADDRESS (C359)

0	1	2	3	4
Public Access	Coop-erator	USGS Only	District Only	Proprietary

MISCELLANEOUS VISIT DATA

RECORD TYPE (C774) V I S T

RECORD SEQUENCE NO. (C737)

DATE OF VISIT (C187) month – day – year

NAME OF PERSON (C188)

MISCELLANEOUS OTHER ID DATA (2 sets shown)

RECORD TYPE (C770) O T I D

RECORD SEQUENCE NO. (C736)

OTHER ID (C190)

ASSIGNER (C191)

RECORD SEQUENCE NO. (C736)

OTHER ID (C190)

ASSIGNER (C191)

MISCELLANEOUS OTHER DATA

RECORD TYPE (C772) O T D T

RECORD SEQUENCE NO. (C312)

OTHER DATA TYPE (C181)

OTHER DATA LOCATION (C182) C Cooperator's Office, D District Office R Reporting Agency Z other

DATA FORMAT (C261) F files, M machine readable, P published, Z other

MISCELLANEOUS LOGS DATA (3 sets shown)

RECORD TYPE (C778) L O G S

RECORD SEQUENCE NO. (C739)

TYPE OF LOG (C199)

BEGINNING DEPTH (C200) .

ENDING DEPTH (C201) .

SOURCE OF DATA (C202) A other gov't D driller G geol-ogist L logs M memory O owner R other reported S reporting agency Z other

DATA FORMAT (C225) F files M machine readable P published Z other

OTHER DATA LOCATION (C226)

RECORD TYPE (C778) L O G S

RECORD SEQUENCE NO. (C739)

TYPE OF LOG (C199)

BEGINNING DEPTH (C200) .

END NG DEPTH (C201) .

SOURCE OF DATA (C202) A other gov't D dr ller G geol-ogist L logs M memory O owner R other reported S reporting agency Z other

DATA FORMAT (C225) F files M machine readable P published Z other

OTHER DATA LOCATION (C226)

RECORD TYPE (C778) L O G S

RECORD SEQUENCE NO. (C739)

TYPE OF LOG (C199)

BEGINNING DEPTH (C200) .

END NG DEPTH (C201) .

SOURCE OF DATA (C202) A other gov't D driller G geol-ogist L logs M memory O owner R other reported S reporting agency Z other

DATA FORMAT (C225) F files M machine readable P published Z other

OTHER DATA LOCATION (C226)

ACOUSTIC LOG:
AS Sonic
AV Acoustic velocity
AW Acoustic waveform
AT Acoustic televiewer

CALIPER LOG:
CP Caliper
CS Caliper, single arm
CT Caliper, hree arm
CM Caliper, multi arm
CA Caliper, acoustic

DRILLING LOG:
DT Drilling time
DR Drillers
DG Geologists
DC Core

ELECTRIC LOG:
EE Electric
ER Single-point resistance
EP Spontaneous potential
EL Long-normal resistivity
ES Short-normal resistivity
EF Focused resistivity
ET Lateral resistivity
EN Microresistivity
EC Microresistivity, forused
EO Microresistivity, lateral
ED Dipmeter

ELECTROMAGNETIC LOG:
MM Magnetic log
MS Magnetic suscep ibiity log
MI Electromagnetic induction log
MD Electromagnetic dual induction log
MR Radar reflection image log
MV Radar direct-wave velocity log
MA Radar direct-wave amplitude log

FLUID LOG:
FC Fluid conductivity
FR Fluid resistivity
FT Fluid temperature
FF Fluid differential temperature
FV Fluid velocity
FS Spinner flowmeter
FH Heat-pulse flowmeter
FE Electromagne ic flowmeter
FD Doppler flowmeter
FA Radioactive tracer
FY Dye tracer
FB Brine tracer

NUCLEAR LOG:
NG Gamma
NS Spectral gamma
NA Gamma-gamma
NN Neutron
NT Neutron activitation
NM Neuclear magnetic resonance

OPTICAL LOG:
OV Video
OF Fisheye video
OS Sidewall video
OT Optical televiewer

COMBINATION LOG:
ZF Gamma, fluid resistivity, temperature
ZI Gamma, electromagnetic induction
ZR Long/short normal resistivity
ZT Fluid resistivity, temperature
ZM Electromagne ic flowmeter, fluid resis ivity, temperature
ZN Long/short normal resistivity, spontaneous potential
ZP Single-point resistance, spontaneous potential
ZE Gamma, long/short normal resis ivity, spontaneous poten ial, single-point resistance, fluid resitivity, temperature

WELL CONSTRUCTION LOG:
WC Casing collar
WD Borehold deviation

OTHER LOG:
OR O her

MISCELLANEOUS NETWORK DATA (3 types shown)

RECORD TYPE (C780) N E T W | RECORD SEQUENCE NO. (C730) | TYPE OF NETWORK (C706) Q W water quality | BEGINNING YEAR (C115) | ENDING YEAR (C116)

TYPE OF ANALYSIS (C120)

A	B	C	D	E	F	G	H	I	J	K	L	M	N	P	Z
physical proper-ties	common ions	trace elements	pesti-cides	nutri-ents	sanitary analysis	codes D&B	codes B&E	codes B&C	codes B&F	codes D&E	codes C,D&E	all or most	codes B&C& radio-active	codes B,C&A	other

SOURCE AGENCY (C117) | [7] FREQUENCY OF COLLECTION (C118) | ANALYZING AGENCY (C307) | [8] PRIMARY NETWORK SITE (C257) | [8] SECONDARY NETWORK SITE (C708)

RECORD TYPE (C780) N E T W | RECORD SEQUENCE NO. (C730) | TYPE OF NETWORK (C706) W L water level | BEGINNING YEAR (C115) | ENDING YEAR (C116)

SOURCE AGENCY (C117) | [7] FREQUENCY OF COLLECTION (C118) | [8] PRIMARY NETWORK SITE (C257) | [8] SECONDARY NETWORK SITE (C708)

RECORD TYPE (C780) N E T W | RECORD SEQUENCE NO. (C730) | TYPE OF NETWORK (C706) W D pumpage or with-drawals | BEGINNING YEAR (C115) | ENDING YEAR (C116)

SOURCE AGENCY (C117) | [7] FREQUENCY OF COLLECTION (C118) | METHOD OF COLLECTION (C133) | [8] PRIMARY NETWORK SITE (C257) | [8] SECONDARY NETWORK SITE (C708)

C	E	M	U	Z
calcu-lated	esti-mated	meter-ed	un-known	other

FOOTNOTES:

[7] FREQUENCY OF COLLECTION CODES

A	B	C	D	F	I	M	O	Q	S	W	Z	2	3	4	5	X
annually	bi monthly	continu-ously	daily	semi-monthly	inter mittent	monthly	one-time only	quarter-ly	semi-annually	weekly	other	bi-annually	every 3 years	every 4 years	every 5 years	every 10 years

[8] NETWORK SITE CODES

1	2	3	4
national,	district,	project,	co-operator,

MISCELLANEOUS REMARKS DATA (4 types shown)

RECORD TYPE (C788) R M K S | RECORD SEQUENCE NO. (C311) | DATE OF REMARK (C184) month – day – year

REMARKS (C185)

Subsequent entries may be used to con inue he remark. Miscellaneous remarks field is limited to 256 characters.

RECORD TYPE (C788) R M K S | RECORD SEQUENCE NO. (C311) | DATE OF REMARK (C184) month – day – year

REMARKS (C185)

Subsequent entries may be used to continue the remark. Miscellaneous remarks field is limited to 256 characters.

DISCHARGE DATA

RECORD SEQUENCE NO. (C147)

DATE DISCHARGE MEASURED (C148) month – day – year

TYPE OF DISCHARGE (C703) P pumped F flow

DISCHARGE (gpm) (C150)

ACCURACY OF DISCHARGE MEASUREMENT (C310)

E	G	F	P
excellent (LT 2%)	good (2%-5%)	fair (5%-8%)	poor (GT 8%)

SOURCE OF DATA (C151)

A	D	G	L	M	O	R	S	Z
other gov't	driller	geologist	logs	memory	owner	other reported	reporting agency	other

METHOD OF DISCHARGE MEASUREMENT (C152)

A	B	C	D	E	F	M	O	P	R	T	U	V	W	X	Z
acoustic meter	bailer	current meter	Doppler meter	estimated	flume	totaling meter	orifice	pitot-tube	reported	trajectory	venturi meter	volumetric meas	weir	unknown	other

PRODUCTION WATER LEVEL (C153)

STATIC WATER LEVEL (C154)

SOURCE OF DATA (C155)

A	D	G	L	M	O	R	S	Z
other gov't	driller	geologist	logs	memory	owner	other reported	reporting agency	other

METHOD OF WATER-LEVEL MEASUREMENT (C156)

A	B	C	D	E	F	G	H	L	M	N	O	P	R	S	T	V	Z
airline	recorder	calibrated airline	differ-ential GP	esti-mated	trans-ducer	pressure gage	calibrated press. gage	geophysi-cal logs	mano-meter	non-rec. gage	observed	acoustic pulse	reported	steel tape	electric tape	calibrated elec. tape	other

PUMPING PERIOD (C157)

SPECIFIC CAPACITY (C272)

DRAWDOWN (C309)

GEOHYDROLOGIC DATA

RECORD TYPE (C748) G E O H

RECORD SEQUENCE NO. (C721)

DEPTH TO TOP OF UNIT (C91)

DEPTH TO BOTTOM OF UNIT (C92)

UNIT IDENTIF ER (C93)

LITHOLOGY (C96)

CONTRIBUTING UNIT (C304)

P	Q	S	N	U
principal aquifer	aggregate of lithologic units	secondary aquifer	no contrib-ution	unknown

LITHOLOGIC MODIF ER (C97)

GEOHYDROLOGIC AQUIFER DATA

RECORD TYPE (C750) A Q F R

RECORD SEQUENCE NO. (C742)

SEQUENCE NO. OF PARENT RECORD (C256)

DATE (C95) month – day – year

STATIC WATER LEVEL (C126)

CONTRIBUTION (C132)

SITE LOCATION SKETCH AND DIRECTIONS

Township ____________ Range ____________

Section # ____________

GWPD 16—Measuring water levels in wells and piezometers by use of a submersible pressure transducer

VERSION: 2010.1

PURPOSE: To make continuous water-level measurements in a well or piezometer by using a submersible pressure transducer.

Materials and Instruments

1. Vented submersible pressure transducer, data logger or data collection platform (DCP), cables, suspension system for the transducer and cables (wire ties or other semipermanent devices), and power supply
2. Data-readout device (i.e., computer loaded with correct software) and data storage modules or other media
3. Locked well cover or recorder shelter and key
4. A water-level tape (steel or electric) graduated in feet, tenths and hundredths of feet, and other materials necessary for depth-to-water measurement
5. Forms including:
    a. Well completion form
    b. Logbook with records of previous measurements for comparison
    c. Transducer calibration worksheet
    d. Water-level measurement field form or groundwater inspection sheet
6. Pencil or pen, blue or black ink. Strikethrough, date, and initial errors; no erasures
7. Calculator
8. Watch
9. Field notebook
10. Spare dessicant
11. Replacement batteries
12. Cleaning supplies for water-level tapes as described in the National Field Manual (Wilde, 2004)
13. Tools including:
    a. High-impedance (digital) multimeter
    b. Connectors
    c. Crimping tool
    d. Contact-burnishing tool or artist's eraser

Data Accuracy and Limitations

1. Water-level measurements for the in-place calibration of pressure transducers should be made to the nearest 0.01 foot.
2. The accuracy of a pressure transducer differs with the manufacturer, measurement range, and depth to water. The measurement error and accuracy standard for most situations are 0.01 foot, 0.1 percent of range in water-level fluctuation, or 0.01 percent of depth to water above or below a measuring point (MP), whichever is least restrictive.
3. Pressure transducers are subject to drift, offset and slippage of the suspension system. For this reason, the transducer readings should be checked against the water level in the well on every visit, and the transducer should be recalibrated periodically and at the completion of monitoring.

Advantages

1. Water levels can be collected at user-defined time scales without making individual manual measurements.
2. Small size allows water levels to be measured in wells or piezometers that are of small diameter, crooked, angled, or that contain pumps or other equipment.
3. The data logger can be left unattended for prolonged periods until data can be downloaded to a portable computer in the field.
4. Some pressure transducers with integrated data loggers are small enough to be placed inside the protective well casing and do not require a separate shelter. Good for high visibility, secure, or below-ground installations.
5. Downloaded data can be imported directly into a spreadsheet or database.
6. Can be interfaced with a DCP to transmit data collected via satellite for near real-time data reporting.
7. Can be installed in a flowing well.

Disadvantages

1. It may be necessary to correct the data for instrument drift, hysteresis, temperature effects, and offsets.
2. Transducers only operate in a limited water-level (pressure) range. The unit must be installed at the appropriate depth in a well so that the water level occurs within the measurement range of the pressure transducer. Wells with a large difference between maximum and minimum water levels may be monitored with reduced resolution using a pressure transducer with a higher range or may require frequent resetting of the depth of the transducer during site visits.
3. Materials in the transducer and cable may react with substances present in the water, causing damage or failure of the instrument.
4. Rapid water-level fluctuations may be missed if they occur between the programmed water-level measurement times.
5. With some data loggers, stored water-level measurements may be lost if the power supply fails.

Assumptions

1. A permanent MP has been established as described in GWPD 3.
2. The user is familiar with the transducer specifications and limitations and has evaluated the required accuracy of the measurements in accordance with the objectives of the study. The transducer's range is appropriate for the range of water levels expected in the observation well (the operating range will not be exceeded).
3. The transducer has been calibrated, either by the manufacturer or by the user, for the conditions expected in the field installation.
4. The transducer is vented to the atmosphere. Data from an absolute transducer must be adjusted to account for changes in atmospheric pressure.
5. If the user is visiting an existing installation, the vent tube is unobstructed, the desiccant is in place, and the well is free of obstructions.

Instructions

This procedure is limited to the installation of vented pressure transducers in observation wells and piezometers for long-term monitoring of water levels (fig. 1). For additional information, and for other applications, see Freeman and others (2004, p. 25–34).

1. If preparing a new installation:
    a. Check that the well is unobstructed. Clear obstructions as described in GWPD 6.
    b. If the well depth is not known, measure the total well depth as described in GWPD 11.
    c. If necessary, install an instrument shelter that will protect the transducer and data logger from vandalism and weather.
    d. Keep the transducer packaged in its original shipping container until it is installed. Connect the transducer, data logger, power supply, and ancillary equipment. Record the model, serial number, and pressure range of the transducer in the field notebook.
    e. Install the pressure transducer by lowering it into the well so that it is submerged below the water surface. Avoid dropping the transducer or permitting sharp contacts with the sides of the well casing. Do not allow the transducer to free fall into the well.

f. Conduct a field calibration of the transducer by raising and lowering it over the anticipated range of water-level fluctuations (Freeman and others, 2004, p. 29). Take three readings at a minimum of five intervals each, during both the raising and lowering of the transducer. Record the data on a calibration worksheet (fig. 2). Calculate a calibration equation for the transducer using the results in figure 2 and a regression equation. If a correction is necessary, apply the correction to the data logger or during post-processing of the water-level record.

g. The transducer should be installed at a point in the well that will not go dry. Estimate the lowest expected water level, and lower the transducer to the desired depth below the water level.

h. Fasten the cable or suspension system to the well head using tie wraps or a weatherproof strain-relief system. If the vent tube is incorporated in the cable, make sure not to pinch the cable too tightly or the vent tube may be obstructed.

i. Make a permanent mark on the cable at the hanging point so that future slippage, if any, can be determined.

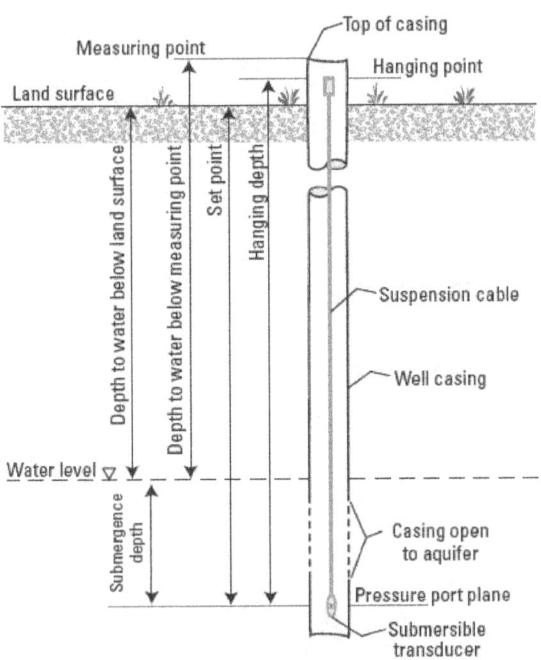

Figure 1. Submersible transducer in an observation well (Freeman and others, 2004, p. 27).

j. Record the well and measuring point (MP) configuration, by drawing a sketch (GWPD 3). Include the MP correction length above the land surface, the hanging point, and the hanging depth (fig. 1).

k. Measure the static water level in the monitor well with a steel (GWPD 1) or electric tape (GWPD 4).

l. Configure the data logger to ensure the channel, scan intervals, and other functions selected are correct. Activate the data logger and set the correct time.

2. If visiting an existing installation:

a. Retrieve groundwater data by using instrument or data logger software.

b. Inspect the equipment to confirm that installation is operating properly. Document the current water level recorded by the sensor (not the most recent water level recorded by the data logger).

c. Measure the depth to water in the well using either GWPD 1 or GWPD 4 to obtain an accurate water-level measurement to compare with the water level measured by the transducer.

d. Record the final water-level measurement on the Inspection of Continuous Record Well field form (fig. 3).

e. If the water-level measurement and transducer reading differ, raise the transducer in the well slightly and take a reading to confirm that the sensor is working. Observe for possible cable kinks or slippage. Return transducer exactly to its original position.

f. Recalibrate the transducer as described in part 1f if necessary (fig. 2).

g. If the water-level measurement differs from the instrumentation reading by an amount specified in the groundwater quality assurance procedures of the local office, record it on the inspection sheet and reset the instrumentation to reflect the proper depth to water.

h. Use the multimeter to check the charge on the battery, and the charging current supply to the battery. Check connections to the data logger, and tighten as necessary. Burnish contacts if corrosion is occurring. Check dessicant. Replace if necessary.

i. Verify the logger channel and scan intervals, document any changes to the data logger program, and reactivate the data logger. Make sure the data logger is operating prior to departure.

U.S. GEOLOGICAL SURVEY
CALIBRATION WORKSHEET FOR SUBMERSIBLE TRANSDUCERS

Data Processing No: ________
Page ______ of ________

Site Name: Official USGS site name ________ Site ID and Number: 8 or 15 digit USGS Site ID ________

M.P used: Nested piezometers often have multiple measuring points Party:

Date (mm/dd/yy): ___ / ___ / ___ Julian: ________ Watch Time: ________ EST CST MST PST Daylight UTC (circle)

Measuring Device: i.e. Calibrated steel tape, calibrated electric tape.

Transducer Information:
Date: ________ Type: ________ Length: ________ Serial No. ________ Output ________

Units of reading: mv, psi, ma Range: i.e.0-5 psi Conversion to Feet: 2.3067 x psi = range of 0 to 11.534 ft.

Calibration marks: Describe what was used to mark the transducer cable for measuring distance moved during the calibration process.

Out-of water reading ________ / ________ Set Point reading: ________ / ________ Scan Rate: ________ Reset? Yes No

Time	Measured Water Level	Cal. Mark	Dist. btwn. Marks	Total Dist.	Readings	
1014	22.35 DBLS		1.00		0.4334 psi	
1015		1		1.00	0.4337	
1016	22.35		1.50		0.4332	
1022	22.35				1.0838 psi	
1023		2		2.50	1.0841	
1024	22.35		1.50		1.0840	
1030	22.34				1.7341 psi	
1031		3		4.00	1.7337	
1032	22.34		1.50		1.7339	
1039	22.33				2.3843	
1040		4		5.50	2.3846	
1041	22.33		1.50		2.3844	
1047	22.33				3.0346	
1048		5		7.00	3.0342	
1049	22.33		1.00		3.0351	
1058	22.32				3.4682	
1059		6		8.00	3.4685	
1100	22.32		1.00		3.4678	
1106	22.32				3.0392	
1107		5		7.00	3.0388	
1108	22.32		1.50		3.0390	
1114	22.32				2.3887	
1115		4		5.50	2.3889	
1116	22.32		1.50		2.3891	
1120	22.31				1.7514	
1121		3		4.00	1.7516	
1122			1.50		1.7517	
1126	22.31				1.1011	
1127		2		2.50	1.1013	
1128	22.31		1.50		1.1010	
1134	22.31				0.4509	
1135		1		1.00	0.4507	WL rise of 0.04
1136	22.31 DBLS		1.00		0.4507	ft. during calib.

Figure 2. Calibration worksheet for submersible transducers (Freeman and others, 2004, p. 30).

INSPECTION OF CONTINUOUS RECORD WELL

Steel Tape or Calibrated Electric Tape Measurement

SITE INFORMATION

SITE ID (C1) [ ]

Measurement Tape ID ______ Date of Field Visit ______

Station name (C12) ______

	1	2	3
Time			
Hold			
Cut			
Tape correction			
WL below MP			
MP correction			
WL below LSD			

Measured by ______

Remarks ______

Barometric Pressure ______ Air Temperature ______

Battery Voltage ______ Replaced? Y / N

Measurement Method: Transducer Float

Checked Float/encoder? Y / N Checked Transducer? Y / N

DATA LOGGER VISIT INFO:

Local time: ______ GMT______ Data logger time: ______

Sensor reading on arrival: ______ Sensor reading on departure: ______ RESET? Y / N

Datum Correction Needed: ______

Retreive data From: ______ To: ______
date/time date/time

Datafile: ______

Remarks: ______

MEASURING POINT DATA (for MP Changes)

M.P. REMARKS (C324)

BEGINNING DATE (C321) [month]-[day]-[year]

ENDING DATE (C322) [ ]-[ ]-[ ]

M.P. HEIGHT (C323) NOTE: (-) for MP below land surface [ ].[ ]

Final Measurement for GWSI

DATE WATER LEVEL MEASURED (C235)	TIME (C709)	STATUS (C238)	METHOD (C239)	TYPE (C243)	WATER LEVEL (C237)
month - day - year					[ ].[ ]

WATER LEVEL TYPE CODE (C243)

L	M	S
below land surface	below meas. pt.	sea level

METHOD OF WATER-LEVEL MEASUREMENT (C239)

A	B	C	E	G	H	L	M	N	R	S (GWPD1)	T	V (GWPD4)	Z
airline,	analog,	calibrated airline,	estimated,	pressure gage,	calibrated press. gage,	geophysical logs,	manometer,	non-rec. gage,	reported,	steel tape,	electric tape,	calibrated elec. tape	other

SITE STATUS FOR WATER LEVEL (C238)

D	E	F	G	H	I	J	M	N	O	P	R	S	T	V	W	X	Z	BLANK
dry,	recently flowing,	flowing,	nearby flowing	nearby recently flowing,	injector site,	injector site monitor,	plugged,	measurement discon.,	obstruction,	pumping,	recently pumped,	nearby pumping,	nearby recently pumped,	foreign substance,	well destroyed,	surface water effects,	other	static

Figure 3. Water-level measurement field form for inspection of continuous record wells. This form, or an equivalent custom-designed form, should be used for continuous recorder inspections and field measurements.

Data Recording

All data times of measurement are recorded in the field notebook or trip log and on the Inspection of Continuous Record Well field form or water-level measurement field form. Depending on the type of data logger used, data from the data logger are transferred to the office computer via field computer or a data module.

References

Cunningham, W.L., and Schalk, C.W., comps., 2011a, Groundwater technical procedures of the U.S. Geological Survey, GWPD 3—Establishing a permanent measuring point and other reference marks: U.S. Geological Survey Techniques and Methods 1–A1, 13 p.

Cunningham, W.L., and Schalk, C.W., comps., 2011b, Groundwater technical procedures of the U.S. Geological Survey, GWPD 4—Measuring water levels by use of an electric tape: U.S. Geological Survey Techniques and Methods 1–A1, 6 p.

Cunningham, W.L., and Schalk, C.W., comps., 2011c, Groundwater technical procedures of the U.S. Geological Survey, GWPD 6—Recognizing and removing debris from a well: U.S. Geological Survey Techniques and Methods 1–A1, 4 p.

Cunningham, W.L., and Schalk, C.W., comps., 2011d, Groundwater technical procedures of the U.S. Geological Survey, GWPD 11—Measuring well depth by use of a graduated steel tape: U.S. Geological Survey Techniques and Methods 1–A1, 10 p.

Cunningham, W.L., and Schalk, C.W., comps., 2011e, Groundwater technical procedures of the U.S. Geological Survey, GWPD 14—Measuring continuous water levels by use of a float-activated recorder: U.S. Geological Survey Techniques and Methods 1–A1, 6 p.

Freeman, L.A., Carpenter, M.C., Rosenberry, D.O., Rousseau, J.P., Unger, Randy, and McLean, J.S., 2004, Use of submersible pressure transducers in water-resources investigations: U.S. Geological Survey Techniques of Water-Resources Investigations, book 8, chap. A3, 50 p.

Lapham, W.W., Wilde, F.D., and Koterba, M.T., 1995, Ground-water data-collection protocols and procedures for the National Water-Quality Assessment Program—Selection, installation, and documentation of wells, and collection of related data: U.S. Geological Survey Open-File Report 95–398, 69 p. + errata.

Wilde, F.D., ed., 2004, Cleaning of equipment for water sampling (version 2.0): U.S. Geological Survey Techniques of Water-Resources Investigations, book 9, chap. A3, section 3.3.8, accessed May 17, 2010, at *http://pubs.water.usgs.gov/twri9A3/*.

GWPD 17—Conducting an instantaneous change in head (slug) test with a mechanical slug and submersible pressure transducer

VERSION: 2010.1

PURPOSE: To obtain data from which an estimate of hydraulic conductivity of an aquifer can be calculated.

During a slug test the water level in a well is changed rapidly, and the rate of water-level response to that change is measured. From these data, an estimate of hydraulic conductivity can be calculated using appropriate analytical methods (for example, Ferris and Knowles, 1963).

A slug test requires a rapid ("instantaneous") water-level change and measurement of the water-level response at high frequency. A rapid change in water level can be induced in many ways, including injecting or withdrawing water, increasing or decreasing air pressure in the well casing, or adding a mechanical device like a plastic rod to displace water. The water-level changes can be measured with many methods, including steel tape, electric tape, air line, wireline/float, and submersible pressure transducers.

One of the most common methods in use is displacement of water with a mechanical slug, measurement of water levels with a submersible pressure transducer, and recording water levels with a data logger. This method combines ease of use, accuracy, and rapidity of water-level measurement. This document describes the mechanical slug/pressure transducer method. This technical procedure can be used with slight modifications if other approaches are used to instantaneously change the water level or measure water-level change.

Materials and Instruments

1. Tools or key to open the well.
2. Field notebook; Pencil or pen, blue or black ink. Strike-through, date, and initial errors; no erasures.
3. Well-construction diagram.
4. Data logger and submersible pressure transducer. A 10-pound-per-square-inch (psi) pressure transducer commonly is used for slug tests because it combines adequate accuracy with an acceptable range of measurement.
5. Slug of polyvinyl chloride (PVC) or other relatively inert material (fig. 1). A slug of solid PVC (fig. 1*C*) is ideal because PVC caps (fig. 1*A*) can catch the well casing during insertion, and PVC plugs (fig. 1*B*) can come loose during the rapid removal of the slug.

   Select the largest diameter and length of slug that will fit in the well without disturbing the transducer. The slug should have a displacement that will provide an adequate change in water level. The slug should displace enough water to provide a measurable change in water level, but not so large as to significantly increase the saturated thickness of the aquifer, disturb the transducer, or affect the speed at which one can raise or lower the slug. A water-level rise between 0.5 and 3 feet (ft) often is adequate. In low permeability formations, a smaller displacement will take less time for full recovery. In high permeability formations (1 to 100 ft per day), a larger displacement is desirable and practical. This usually can be generated with a slug diameter about 1 inch less than the well diameter and a length of 3 ft or more (lengths greater than 5 ft are awkward to handle in the field). Tables 1 and 2, respectively, provide theoretical displacement volumes for various slugs and volumes necessary for specific water-level changes.
6. Nylon cord or other strong line of sufficient length to reach below the water level in order to secure the slug.
7. Wooden rod, or 2 by 4 to secure the slug line.
8. Tripod or other device to support the slug line (optional).
9. Bungee cord or other device to secure the transducer cable and support line.
10. Water level measuring device (steel or electric tape).
11. Appropriate decontamination equipment, if necessary.
12. Field computer (optional).
13. Stopwatch (optional).

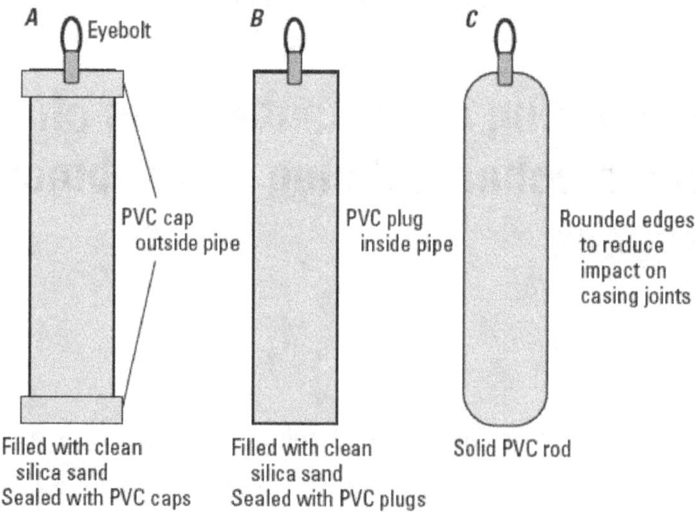

Figure 1. Polyvinyl chloride (PVC) plastic slug. *A*, Solid 2-inch PVC pipe with external cap. *B*, Solid 2-inch PVC pipe with internal plug. *C*, Solid 2-inch PVC rod.

Table 1. Slug displacement volume, in cubic feet, for a specific slug diameter and length.

Slug length (feet)	Slug diameter (inches)						
	1	1.5	2	2.5	3	3.5	4
2	0.011	0.025	0.044	0.068	0.098	0.134	0.175
3	0.016	0.037	0.065	0.102	0.147	0.200	0.262
4	0.022	0.049	0.087	0.136	0.196	0.267	0.349
5	0.027	0.061	0.109	0.170	0.245	0.334	0.436
6	0.033	0.074	0.131	0.205	0.295	0.401	0.524

Table 2. Volume of water, in cubic feet, required to raise the water level a prescribed distance within a specific well diameter.

Well diameter (inches)	0.3-foot rise	0.5-foot rise	1-foot rise	1.5-foot rise	2-foot rise	3-foot rise
2	0.007	0.011	0.022	0.033	0.044	0.065
3	0.015	0.025	0.049	0.074	0.098	0.147
4	0.026	0.044	0.087	0.131	0.175	0.262
6	0.059	0.098	0.196	0.295	0.393	0.589
8	0.105	0.175	0.349	0.524	0.698	1.047
10	0.164	0.273	0.545	0.818	1.091	1.636

Data Accuracy and Limitations

1. The accuracy of a slug test is a function of many factors, including well construction, field procedures, and analysis method. Rapidly changing the water level in a well can be done by submerging an object (slug) in the water, causing the water level to rise instantaneously. Displaced water will move from the well to the geologic formation until the hydraulic head falls to the original static or equilibrium level. This is called a falling head test or "slug in test." After the water level reaches equilibrium, quickly removing the slug causes the water level to fall instantaneously. Water will move from the formation into the well until the hydraulic head returns to the equilibrium level. This is called a rising head test, "slug-out test," or bailer test. Because the early-time data for these tests are most important for the subsequent analysis, the data logger should begin collecting data just before the slug is submerged or removed from the well. The initial time can be adjusted during analysis, but the logger must be collecting data at a frequency of at least several samples per second when the water level begins to change. After the first minute or two of data collection, the sampling interval can be increased. Data loggers designed for aquifer tests and slug tests frequently have internal programs that allow for rapid data collection at early time and gradual increase of the sampling interval over time (a logarithmic time scale).

2. Some transducers have more rapid recording rates than others. If the slug test is being done in a formation of high hydraulic conductivity, select a transducer that can transmit at very small time increments (tenths of a second).

3. Due to the accuracy limitations of slug tests, results should be reported to one significant figure.

Advantages

1. Potentially contaminated water requiring special disposal is not removed from the well.

2. The slug test can be conducted quickly and is therefore relatively inexpensive.

3. Only one well is needed for the test (no need for other observation wells), and a pump is not required.

4. Because the slug-test data to be analyzed for an estimate of hydraulic conductivity are collected within a few minutes of the test initiation, this technique can be used near pumped wells or where well interference is expected, as long as the expected water-level changes occur slowly in comparison to the time for which the slug-test data will be analyzed.

Disadvantages

1. The collected data represent only a small volume of aquifer material near the tested well.

2. The test may be influenced by the well filter pack, skin effects, or poor well development.

Assumptions

1. Operator is familiar with the operation of data loggers and submersible pressure transducers. The data logger/transducer can measure and record at a high frequency (less than or equal to one second in highly transmissive formations).

2. The well is free of obstructions which might hinder water-level measurement or introduction or removal of the mechanical slug.

3. The water level is easily accessible from the surface (within approximately 100 ft) and is within the length of the transducer cable.

4. Column of water in the well is long enough to cover the transducer and the slug.

5. The well is properly constructed and developed.

6. Well construction details such as well depth, screen length, borehole radius, filter pack, and well radius are known.

7. The hydraulic conductivity of the aquifer is not extremely low. A slug test is an acceptable method in low-permeability formations, but a transducer may not be necessary in this situation. The water level in the well should recover within minutes or hours for this procedure.

Instructions

1. Confirm well identification with well-construction diagram.

2. Measure the total depth of the well (see GWPD 11).

3. Measure the water level in the well (see GWPD 1 or GWPD 4). This should be repeated at the end of the test for long duration slug tests. The column of water in the well should be long enough to cover the transducer and the slug.

4. Document the static water level, well diameter, well depth, and screened interval in field notebook. The diameter of the hole, nature of filter pack, and type of screen also are documented, if known.

5. Place the transducer in the well below the level at which the slug will be submerged, but not so low that the range of transducer might be exceeded at the highest anticipated water level. Secure the transducer in place. The transducer should not move during the test.

6. Measure (estimate) the maximum length of slug line that will be used. This length should allow the slug to completely submerge, about 1 ft below water surface.

7. Allow the transducer to adjust to the new pressure and temperature following manufacturer's guidance. This also provides time for the water level to recover prior to the test.

8. If needed, set up a tripod or some other device from which the slug can be lowered and raised in the well. Lower the clean, decontaminated slug to a point just above the water level and secure it in place. Take care not to move or kink the transducer line (fig. 2*A*). A simple approach of securing the slug is to tie a loop of cord that would hold the slug about 1 ft above the water surface and then tie off a second loop at the length of cord required for the entire slug to submerge. Put both of these loops over a rod or a wooden 2 by 4 that can rest across the top of the well casing.

9. Prepare the data logger. The data logger should be set to record data as frequently as possible during the first minutes of the test, and it can be set to record less frequently during later time. Recording in seconds on a logarithmic time scale meets this objective.

10. Establish a starting water level for the transducer and data logger. Data analysis is based on the change in water level rather than a comparison to a standard datum. The transducer starting water level can be set to zero, a value equal to the head of water above the transducer, or any other value.

Slug In Test

11. Begin the test by starting the data logger and nearly simultaneously submerging the slug quickly but gently into the water to minimize disturbance at the water surface or movement of the transducer cable (fig. 2*B*). Secure the slug cord to the wooden rod to maintain its position below the water level.

12. After 1 minute and periodically thereafter, check the status of the water-level reading with the data logger/transducer or with a water-level measuring tape.

13. When the water level is equal to the initial water level, or when readings change less than 0.01 ft per 10 minutes, stop the test. This is the end of the falling head, or slug in test. You are now ready to begin the rising head, or slug out test.

Slug Out Test

14. Establish a starting water level for the transducer and data logger. Data analysis is based on the change in water level rather than a comparison to a standard datum. The transducer starting water level can be set to zero, a value equal to the head of water above the transducer, or any other value.

15. Prepare the data logger. The data logger should be set to record data as frequently as possible during the first minutes of the test, and it can be set to record less frequently during later time. Recording in seconds on a logarithmic time scale meets this objective.

16. Begin the test by starting the data logger and nearly simultaneously withdrawing the slug quickly but gently from the water to minimize disturbance at the water surface or movement of the transducer cable. The slug need not be withdrawn completely out of the well, but should

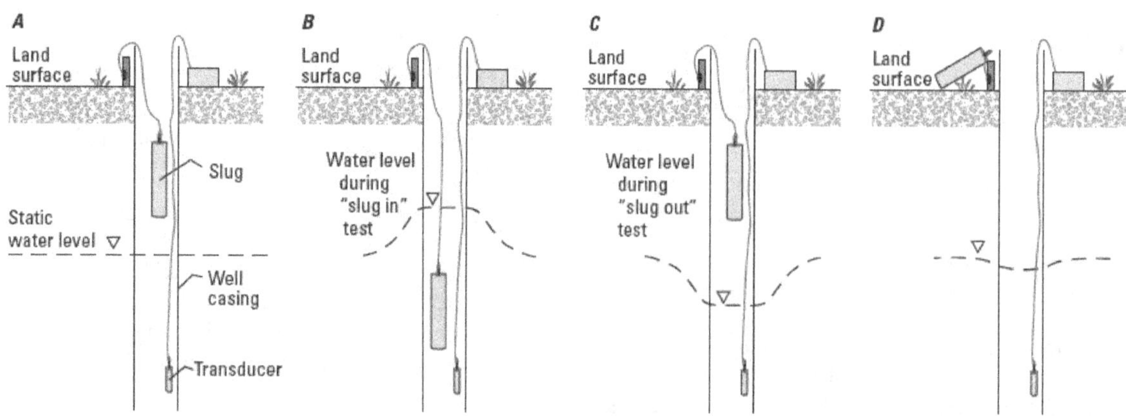

Figure 2. Well diagram with polyvinyl chloride (PVC) plastic slug (*A*) poised just above the water level for falling head or slug in test, (*B*) submerged below the water level for falling head or slug in test, (*C*) removed just above the water level for rising head or slug out test, and (*D*) removed from the well for rising head or slug out test.

be out of the water (fig. 2*C* or 2*D*). Secure the slug cord to the wooden rod to maintain its position above the water level.

17. After 1 minute and periodically thereafter, check the status of the water-level reading with the data logger/transducer or with a measuring tape.

18. When the water level is equal to the initial water level, or when readings change less than 0.01 ft per 10 minutes, stop the test. This is the end of the rising head, or slug out test.

19. Review the data for completeness and accuracy. This can be done on the data logger or on a field computer (preferred). Optionally, the test can be analyzed in the field on a field computer using aquifer test software.

20. Repeat the entire procedure at least once as time permits, so two complete sets of falling and rising head test data are collected (four tests).

Data Recording

1. All calibration and maintenance data associated with the data logger, steel or electric tape, and submersible pressure transducer are recorded in calibration and maintenance equipment logbooks.

2. Complete a field report with date, time, well identifier, type of test (rising or falling head), composition and dimensions (or volume) of the slug, and the name of data files. (Use site ID or well name, date, and year in the file name: for example, 424531077564201.19960101, or Well8.19960101.)

3. Data are downloaded to an office computer for processing. Results are interpreted and submitted for Bureau approval. Original data are stored in the office aquifer test archive, and result is recorded on the Groundwater Site Inventory form (fig. 3, Form 9-1904-D1).

FORM NO. 9-1904-D1
Revised January 2010, NWIS 4.9

Coded by ____________________ File Code ____________________
Checked by ____________________ Date ____________________
Entered by ____________________ Regional approval date ____________________

U.S DEPT. OF THE INTERIOR
GEOLOGICAL SURVEY

GROUNDWATER SITE INVENTORY
Hydraulics Data

AGENCY CODE (C4) [| | | |] SITE ID (C1) [| | | | | | | | | | | | | |]

RECORD TYPE (C744) H Y D R RECORD SEQUENCE NO. (C790) [| |]

HYDRAULIC UNIT IDENTIFIER (C100) [| | | | | | |] DEPTH TO TOP OF INTERVAL (C101) [| | |] . [|] DEPTH TO BOTTOM OF INTERVAL (C102) [| | |] . [|]

HYDRAULICS UNIT TYPE (C103) A C
aquifer confining unit

REMARKS - Method of determining hydraulics data (C104)

HYDRAULICS SOURCE AGENCY (C305) [| | | |] WEB-READY FLAG (C874) Y C P L
ready to display, condi- tional, propri- etary, local use only

---

RECORD TYPE (C746) C O E F SEQUENCE NO. OF PARENT RECORD (C99) [| |] RECORD SEQUENCE NO. (C106) [| |]

TRANSMISSIVITY(C107) [| | | | | | |]

HORIZONTAL CONDUCTIVITY (C108) [| | | | | | | |] . [| |] VERTICAL CONDUCTIVITY (C109) [| | | | |] . [| | | |]

STORAGE COEFFICIENT (C110) . [| | | | | | |] LEAKANCE (C111) [| | |] . [| | |]

DIFFUSIVITY (C112) [| | | | | | | | | | | | |] SPECIFIC STORAGE (C113) . [| | | | | | | | |]

BAROMETRIC EFFICIENCY (Percent) (C271) [| |] POROSITY (C306) . [| |]

WEB-READY FLAG (C875) Y C P L
ready to display, condi- tional, propri- etary, local use only

Figure 3. Groundwater Site Inventory for Hydraulics Data, Form 9-1404-D1.

Procedures References

Cunningham, W.L., and Schalk, C.W., comps., 2011a, Groundwater technical procedures of the U.S. Geological Survey, GWPD 1—Measuring water levels by use of a graduated steel tape: U.S. Geological Survey Techniques and Methods 1–A1, 4 p.

Cunningham, W.L., and Schalk, C.W., comps., 2011b, Groundwater technical procedures of the U.S. Geological Survey, GWPD 3—Establishing a permanent measuring point and other reference marks: U.S. Geological Survey Techniques and Methods 1–A1, 13 p.

Cunningham, W.L., and Schalk, C.W., comps., 2011c, Groundwater technical procedures of the U.S. Geological Survey, GWPD 4—Measuring water levels by use of an electric tape: U.S. Geological Survey Techniques and Methods 1–A1, 6 p.

Cunningham, W.L., and Schalk, C.W., comps., 2011d, Groundwater technical procedures of the U.S. Geological Survey, GWPD 11—Measuring well depth by use of a graduated steel tape: U.S. Geological Survey Techniques and Methods 1–A1, 10 p.

Method References

American Society for Testing of Materials, 1991, ASTM Method D4044-91: Philadelphia, Pennsylvania, American Society for Testing of Materials.

Ferris, J.G., and Knowles, D.B., 1963, The slug-injection test for estimating the coefficient of transmissibility of an aquifer, *in* Bentall, Ray, comp., Methods of determining permeability, transmissibility, and drawdown: U.S. Geological Survey Water-Supply Paper 1536–I, p. 299–304.

Hoopes, B.C., ed., 2004, User's manual for the National Water Information System of the U.S. Geological Survey, Ground-Water Site-Inventory System (version 4.4): U.S. Geological Survey Open-File Report 2005–1251, 274 p.

Analysis References

Bouwer, Herman, 1989, The Bouwer and Rice slug test—An update: Ground Water, v. 27, no. 3, p. 304–309.

Bouwer, Herman, and Rice, R.C., 1976, A slug test method for determining hydraulic conductivity of unconfined aquifers with completely or partially penetrating wells: Water Resources Research, v. 12, no. 3, p. 423–428.

Butler, J.J., Jr., 1997, The design, performance, and analysis of slug tests: Boca Raton, Florida, Lewis Publishers, 252 p.

Cooper, H.H., Bredehoeft, J.D., and Papodopulos, S.S., 1967, Response of a finite-diameter well to an instantaneous charge of water: Water Resources Research, v. 3, no. 1, p. 263–269.

Dawson, K.J., and Istok, J.D., 1991, Aquifer testing—Design and analysis of pumping and slug tests: Chelsea, Michigan, Lewis Publishers, 344 p.

Halford, K.J., and Kuniansky, E.L., 2002, Documentation of spreadsheets for the analysis of aquifer-test and slug-test data: U.S. Geological Survey Open-File Report 02–197, 54 p. (Also available at *http://pubs.usgs.gov/of/2002/ofr02197/*.)

Hvorslev, M.J., 1951, Time lag and soil permeability in ground-water observations: Vicksburg, Mississippi, U.S. Army Corps of Engineers, Waterways Experiment Station, Bulletin No. 36, p. 1–50.

HydroSOLVE, Inc., 1998, AQTESOLV for Windows User's Guide: Reston, Virginia, HydroSOLVE, 128 p.

Krusman, G.P., and deRidder, N.A., 1990, Analysis and evaluation of pumping test data (2d ed.): Wageningen, The Netherlands, International Institute for Land Reclamation and Improvement, 377 p.

www.ingramcontent.com/pod-product-compliance
Lightning Source LLC
Chambersburg PA
CBHW081452170526
45166CB00008B/2401

* 9 7 8 1 5 0 0 2 1 9 7 5 8 *